NONLINEAR OPTICS

NONLINEAR OPTICS

4th Edition

Nicolaas Bloembergen
Harvard University

World Scientific
Singapore • New Jersey • London • Hong Kong

Published by

World Scientific Publishing Co. Pte. Ltd.

P O Box 128, Farrer Road, Singapore 912805

USA office: Suite 1B, 1060 Main Street, River Edge, NJ 07661

UK office: 57 Shelton Street, Covent Garden, London WC2H 9HE

British Library Cataloguing-in-Publication Data
A catalogue record for this book is available from the British Library.

4th Edition 1996 by World Scientific Publishing Co. Pte. Ltd.
2nd printing 1998

NONLINEAR OPTICS — 4TH EDITION

Original Edition 1965 published by W. A. Benjamin, Inc.
Third Printing 1992 by Addision–Wesley Publishing Co., Inc.

ISBN 981-02-2599-7 (pbk)

Printed in Singapore by Uto-Print.

About the Author

Nicolaas Bloembergen was born in 1920 in the Netherlands. He obtained the equivalent of a bachelor's and master's degree in physics at the University of Utrecht. He carried out research for the PhD degree under the guidance of Professor E M Purcell at Harvard University in 1946 and 1947, later becoming a postdoctoral research fellow with Professor C J Gorter at the Kamerlingh Onnes Laboratory at the University of Leiden, where he obtained his PhD in 1948. From 1949 to 1951, he was a Junior Fellow in the Society of Fellows at Harvard University, where he has held a series of tenured faculty positions since 1951. He is now Gerhard Gade University Professor Emeritus.

In 1981, Professor Bloembergen received the Nobel Prize in Physics. He was also the recipient of the Lorentz medal of the Royal Dutch Academy in 1978. The President of the United States of America has also awarded him the national Medal of Science. His other awards include the Buckley Prize of the American Physical Society, the Ballantine Medal of the Franklin Institute, the Ives Medal of the Optical Society of America, and the Medal of Honor of the Institute of Electrical and Electronic Engineers.

Professor Bloembergen is a member of the US National Academy of Sciences, the American Academy of Arts and Sciences, as well as the American Philosophical Society. He is a foreign associate of academics in the Netherlands, France, Germany, India ,and Norway.

About the Author

Preface

This monograph was written in the spring of 1964 in preparation for a series of lectures presented at the 1964 Les Houches summer school. The text was originally published by W.A. Benjamin, Inc., as part of the *Frontiers in Physics* series, edited by David Pines. It was subsequently reprinted several times by Addison-Wesley Publishing Company, Inc., most recently as part of its *Advanced Book Classics* series.

It is gratifying that there is a continuing demand for this presentation of basic physical ideas which retain their validity and relevance. The field of Nonlinear Optics today has grown into a vast enterprise with a considerable potential for technological applications. New nonlinear optical materials and devices are in various stages of development. Purely optical information processing looms on the horizon. At the same time, basic research in nonlinear optical phenomena retains its vitality. Topics of current interest include, among others, optical solitons, femto-second time-resolved spectroscopy and squeezed quantum states.

Advanced scientists and engineers, as well as students entering the vast and burgeoning field of Nonlinear Optics, may be interested in the basic physical ideas and early historical developments described in this monograph. No attempt has been made to revise this material dating from 1964. An epilogue provides some corrective notes as well as a list of more up-to-date textbooks.

N. Bloembergen

Cambridge, Massachusetts
November 1995

Preface
to the Original Edition

This monograph is based on lectures, prepared for a course on quantum electronics at Harvard University in the spring of 1963 and for the summer school in Les Houches in 1964. The field of nonlinear optics is quite young. It deals with phenomena that occur at very high light intensities obtainable in laser beams. It represents one of the most interesting fields of research made possible by the development of powerful lasers.

It is perhaps foolhardy to write a monograph about nonlinear optics at this time, when new results are still announced at a high rate in scientific journals. It could be argued that such a monograph would at best contribute to its own rapid obsolescence. Nevertheless, it is hoped that it may have some more lasting value. The general principles of Maxwell's electromagnetic theory and of quantum mechanics are well established. Their domain of application is extended to include higher order interactions between light and matter in terms of nonlinear susceptibilities.

The nonlinear response of circuit elements at audio-radio and microwave frequencies is well known to the electrical engineer. In this monograph the analogous phenomena at optical frequencies are discussed. The concepts of harmonic generation, parametric amplification, modulation and rectification all have their counterparts in the visible region of the electromagnetic spectrum. The material is organized so that a pure classical description can be followed by those who have a knowledge of electromagnetic theory but are not familiar with quantum mechanics. They may skip Chapter 2 in which the quantum theory of linear and nonlinear susceptibilities is treated. This volume is intended for all who have an active interest in the field of quantum electronics, whether they are physicists interested in nonlinear electromagnetic properties of matter, electrical engineers interested in communications or high power applications at visible frequencies, or optical scientists interested in the behavior of light rays at very high intensities.

Since the field of nonlinear optics is still in a stage of rapid expansion, no effort has been made to give a complete bibliography nor to achieve a complete coverage of all experimental data. The fundamental theoretical ideas and the basic experimental results are emphasized.

The author is indebted to Dr. J. Ducuing, Dr. Y.R. Shen, and Dr. D. Forster who have carefully read the manuscript and suggested many corrections. Any errors that remain are entirely the responsibility of the author. The permission of the edits of *The Physical Review* and the respective coauthors to reproduce the three appendices is gratefully acknowledged. The author is indebted to Drs. P.S. Pershan, and R.W. Terhune for making available some material before publication. The author wishes to express his thanks to Elizabeth Dixon who typed the entire manuscript on a tight schedule.

This monograph is dedicated to Deli Bloembergen, whose encouragement and understanding were a decisive factor in its timely completion.

N. Bloembergen

Cambridge, Massachusetts
July 1964

Contents

NONLINEAR OPTICS

1

CLASSICAL INTRODUCTION

1-1 NONLINEAR SUSCEPTIBILITIES

Nonlinear properties of Maxwell's constitutive relations,

$$D = \epsilon(E)E \qquad B = \mu(H)H \tag{1-1}$$

have been known from the beginning. The dielectric constant and magnetic permeability can be functions of the field strengths. The nonlinear permeability of ferromagnetic media was of prime concern in the design of electrical machinery in the nineteenth century. This nonlinearity has its origin in domain wall motion and domain rotation. It can be a source of harmonic distortion in audio-amplifiers using inductances with ferromagnetic cores. Paramagnetic saturation described by Langevin and Brillouin may be considered as a nonlinearity at zero frequency. Magnetic and dielectric amplifiers are based on the nonlinearity of ferromagnetica and ferroelectrica at relatively low frequencies. The nonlinear electromagnetic response of a plasma has been known for a long time. It was invoked to explain the "Luxembourg" effect in ionospheric propagation of radiowaves. More recently, the nonlinear properties of plasmas and other materials have been investigated at microwave frequencies. The nonlinearity in ferromagnetic resonance, for example, has been used to generate second and higher harmonics in the microwave region of the electromagnetic spectrum.

Nonlinear properties in optical region have been strikingly demonstrated by harmonic generation of light. Franken[1] and coworkers detected ultraviolet light (λ = 3470 Å) at twice the frequency of a ruby laser beam (λ = 6940 A) when this beam traversed a quartz crystal. A schematic diagram of the experimental arrangement

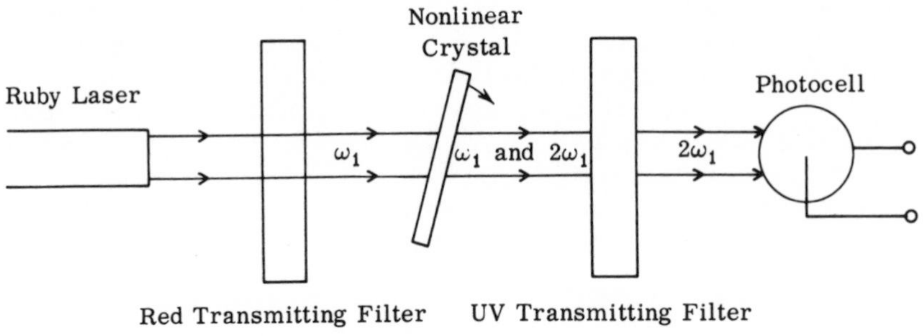

Figure 1-1. Experimental arrangement for the detection of second harmonic generation of light.

is shown in Figure 1-1. This experiment, carried out in 1961, marked the beginning of a large activity in both experimental and theoretical nonlinear optical properties. This work has been reviewed in a number of papers,[2,3,4] where the reader may find rather complete references to the literature. In this volume no attempt will be made to give a comprehensive review of all the detailed work in this field of endeavor, but emphasis will be placed on the development and discussion of the diverse phenomena from a unified point of view. The nonlinear material properties are described by expanding the polarization in a power series in the field. For the pure electric dipole case one has, for example,

$$\mathbf{P} = \chi \cdot \mathbf{E} + \chi : \mathbf{EE} + \chi \vdots \mathbf{EEE} + \cdots \tag{1-2}$$

The first term defines the usual linear susceptibility, the second term, the lowest order nonlinear susceptibility, and so on. This procedure is useful because the optical nonlinearities are small. In fact, their very small size is largely responsible for their late experimental discovery, which had to wait for the development of powerful lasers. Large electric fields with peak amplitudes of about one million volts/cm, corresponding to a flux density of about 10^9 watts/cm^2, are now available at optical frequencies from Q-switched lasers.[5] Effects are under favorable conditions already detectable with a flux density of less than one mw/cm^2.

The theory of wave propagation in nonlinear media can be developed along purely classical lines and the nonlinear optical properties can be discussed along the same lines as the linear optical properties at the turn of the last century. Maxwell, Hertz, Lorentz, and Drude, however, lacked the stimulation of experimental findings. The advent of stimulated emission of light has suddenly changed this. The behavior of light beams at high intensity is studied vigorously.

Many generalizations of the classical laws of optics to the regime of intensities, where nonlinearities are important, have already been made. At the same time the nonlinear susceptibilities themselves are of intrinsic interest in the study of the structure of matter. This situation is of course similar in the linear case. The linear index of refraction determines the paths of light rays. Conversely, knowledge of its behavior gives information about the nature of the material.

1-2 CLASSICAL ATOMIC MODELS OF NONLINEARITY

The Free Electron Gas

Consider the motion of a single electron in a plasma under the influence of a linearly polarized light wave:

$$\begin{aligned} B_y = E_x &= E \exp(ikz - i\omega t) + E^* \exp(-ikz + i\omega t) \\ &= 2 \operatorname{Re}\{E \exp(ikz - i\omega t)\} \text{ with } k = \omega c^{-1} \end{aligned} \tag{1-3}$$

Note that the real amplitude of the wave is $2|E|$, or our amplitude $|E|$ is half that of the usual definition. This new convention, introduced by Pershan,[4] leaves the linear susceptibility unchanged, but increases the lowest order nonlinear susceptibility by a factor of two. Additional factors of two appear in higher order susceptibilities. The new convention has a definite advantage in the discussion of certain symmetry properties of nonlinear susceptibilities. The limiting cases that two frequencies become equal, or one of the frequencies goes to zero, have been a source of confusion.[6] The new system largely avoids difficulties with factors two during the computation. One may revert to the old convention at the end of the calculation. The Fourier component of a physical quantity at the frequency $+\omega$ has a time dependence $\exp(-i\omega t)$, the Fourier component at $-\omega$ goes as $\exp(+i\omega t)$.

The equations of motion for a single electron in the plasma are,

$$\begin{aligned} m\ddot{x} &= eE_x - ec^{-1}\dot{z}B_y - m\dot{x}/\tau \\ m\ddot{y} &= -m\dot{y}/\tau \\ m\ddot{z} &= ec^{-1}\dot{x}B_y - m\dot{z}/\tau \end{aligned} \tag{1-4}$$

The phenomenological collision time τ describes the damping of the motion in a statistical sense and insures a steady state response independent of the initial conditions. The Lorentz force gives rise to harmonics. A steady state solution of Eqs. (1-3) and (1-4) can be found by successive approximation in the form of a Fourier series.

In the first or linear approximation one has the well-known result

$$x(\omega) = \frac{-eE\exp(ikz - i\omega t)}{m(\omega^2 + i\omega\tau^{-1})} \tag{1-5}$$

If this linear solution is substituted into the last equation of (1-4), the lowest order nonlinear approximation gives

$$z(2\omega) = \frac{-ie^2E^2\exp(2ikz - 2i\omega t)}{m^2c(4\omega + 2i\tau^{-1})(\omega^2 + i\omega\tau^{-1})} \tag{1-6}$$

The linear dipole moment $ex(\omega)$ gives rise to the well-known Thomson formula, describing the Rayleigh scattering of light by free electrons. In a similar way the dipole moment $ez(2\omega)$ radiates at the second harmonic frequency. This nonlinear scattering process may roughly be described by saying that two incident quanta are taken away from the incident beam and one quantum at twice the frequency is radiated with the intensity pattern of a dipole oriented along the direction of the incident beam.

The incoherent nonlinear scattering by individual electrons is, however, not of practical interest. The nonlinear phenomena are sufficiently weak that only the coherent radiation of a large assembly of particles is detectable. A similar distinction between coherent and incoherent scattering holds for the linear properties of a plasma. The attention should be focused on the average polarization in a small volume and the index of refraction of the plasma, rather than on the incoherent Rayleigh scattering due to fluctuations in density. In liquids and crystals the relative density fluctuations are even smaller and the main interest is in the coherent polarization of the medium.

If the average density of electrons in the plasma is N_0 per cm^3, the polarization is

$$P_x(\omega) = \chi(\omega)E_x(\omega) = N_0ex(\omega). \tag{1-7}$$

Since at optical frequencies $\omega\tau \ll 1$, Eqs. (1-5) and (1-7) give immediately the familiar result for the susceptibility of a plasma:

$$\epsilon - 1 = 4\pi\chi = -4\pi N_0e^2/m\omega^2$$

In a similar way, the nonlinear polarization at the second harmonic frequency is given by

$$P_z(2\omega) = N_0ez(2\omega).$$

There is no coherent radiation at 2ω in an infinite plasma, because this polarization is parallel to the direction of propagation. Coherent second harmonic radiation is possible at the boundary of a plasma.

There is also a small dc current in the longitudinal direction in this approximation, generated by the term $\dot{x}(\omega)B_y(-\omega) + \dot{x}(-\omega)B_y(+\omega)$. One finds the dc current density

$$J_z(o) = N_o e\dot{z}(o) = \frac{iN_o e^3 \omega\tau |E|^2}{m^2 c(\omega^2 + i\omega\tau^{-1})} \tag{1-8}$$

This term represents the classical analogue to momentum transfer in Compton scattering.

The procedure can of course readily be extended to higher harmonics. There is a component of polarization $P_x(3\omega)$ which will generate third-harmonic radiation.

The nonlinearities in plasmas at microwave frequencies can be quite large, especially in the presence of a dc magnetic field. Then cyclotron resonance effects can occur.[7] In actual plasmas at lower frequencies other nonlinearities, besides the Lorentz force, are important. Hydrodynamic pressure gradients and induced variations in the electron density must be considered. A rather complete discussion of all these effects has been given by Whitmer and Barrett.[8]

The Anharmonic Oscillator

A very useful model used by Drude and Lorentz[9] to calculate the linear polarization of a medium describes the electrons as harmonically bound particles. The resonant frequencies of the oscillators were taken to correspond to the observed atomic spectral lines. Actually the valence electrons are bound by the Coulomb field of the ion cores. For very large deviations from equilibrium the anharmonicity of the electron oscillators must be taken into account. Such a model had already been used by Rayleigh to explain nonlinearities in acoustic resonators.[10] Consider therefore the motion of a one-dimensional anharmonic oscillator with damping, driven by an electric field with Fourier components at the frequencies $\pm\omega_1$ and $\pm\omega_2$.

$$\ddot{x} + \Gamma\dot{x} + \omega_0^2 x + vx^2 = (2e/m)\,\mathrm{Re}\{E_1 \exp(ik_1 z - i\omega_1 t) + E_2 \exp(ik_2 z - i\omega_2 t\} \tag{1-9}$$

The linear approximation gives immediately the well-known result

$$x(\omega_1) = \frac{e}{m(-\omega_1^2 + \omega_0^2 - i\omega_1\Gamma)} E_1 \exp(ik_1 z - i\omega_1 t) \tag{1-10}$$

In the lowest order nonlinear approximation one finds terms at the second-harmonic frequencies $2\omega_1$, $2\omega_2$, a term at zero frequency representing the rectification of light by the quadratic nonlinearity vx^2, and terms at the sum and difference beats between the two light waves, $\omega_1 + \omega_2$ and $\omega_1 - \omega_2$. Only two Fourier components will be reproduced here, a second harmonic

$$x(2\omega_1) = \frac{-(e^2/m^2)vE_1^2 \exp(2ik_1 z - 2i\omega_1 t)}{D^2(\omega_1)\, D(2\omega_1)} \tag{1-11}$$

and the difference beat

$$x(\omega_1 - \omega_2) = \frac{-(e^2/m^2)vE_1E_2^* \exp[i(k_1 - k_2)z - i(\omega_1 - \omega_2)t]}{D(\omega_1)\, D^*(\omega_2)\, D(\omega_1 - \omega_2)} \tag{1-12}$$

where the abbreviated notation

$$D(\omega) = \omega_0^2 - \omega^2 - i\Gamma\omega = D^*(-\omega)$$

for the denominators has been introduced. From Eq. (1-11) one derives immediately a nonlinear polarization and a nonlinear susceptibility describing second-harmonic production

$$P_x^{NL}(2\omega) = \chi_{xxx}(2\omega,\omega,\omega)E_x^2(\omega) \tag{1-13}$$

$$\chi_{xxx}(2\omega,\omega,\omega) = \frac{-N_0(e^3/m^2)v}{D^2(\omega)\, D(2\omega)} \tag{1-14}$$

The indices xxx are of course superfluous in the one-dimensional example. They serve as a reminder of the fact that the susceptibility really connects three vectors and has the transformation properties of a third-rank tensor. The extension to a three-dimensional harmonic oscillator is straightforward. A similar expression can of course be derived for the nonlinear susceptibility

$$\chi_{xxx}(\omega_1 - \omega_2,\ \omega_1,\ -\omega_2)$$

Note that the dispersion of the lowest order nonlinear susceptibility is described with reference to a set of three frequencies, rather than a single frequency. The dispersive properties are enhanced near resonance of one of the denominators. If, for example, the difference frequency is equal to the resonant frequency, $D(\omega_1 - \omega_2) = i\omega_0\Gamma$ for $\omega_1 - \omega_2 = \omega_0$, the nonlinear susceptibility at the difference frequency is far larger than all others.

When $\omega_1 - \omega_2$ is equal or nearly equal to the resonant frequency ω_0, it is permissible to retain only the Fourier component (1-12) in the calculation of next higher order nonlinearities. It beats in vx^2 with linear terms to yield components at $2\omega_1 - \omega_2$, $\omega_1 - 2\omega_2$, in addition to terms at the original frequencies ω_1 and $-\omega_2$. In this manner one finds, for example,

$$x^{NL}(\omega_2)^* = x^{NL}(-\omega_2) = \frac{(e^3/m^3)v^2}{(D^*(\omega_2))^2 |D(\omega_1)|^2 D(\omega_1 - \omega_2)} E_2^* |E_1|^2$$

or

$$\chi_{xxxx}(\omega_2 = \omega_2 + \omega_1 - \omega_1) = \frac{N_0(e^4/m^3)v^2}{D^2(\omega_2)|D(\omega_1)|^2 D^*(\omega_1 - \omega_2)} \tag{1-15}$$

At resonance $\omega_1 - \omega_2 = \omega_0$, and $\omega_1 \gg \omega_0$, $D(\omega_1 - \omega_2)$ is purely imaginary; the other factors in the denominator are real and

$$\chi_{xxxx}(\omega_2 = \omega_2 = +\omega_1 - \omega_1) = -\frac{iN_0(e^4/m^3)v^2}{\omega_0 \Gamma(\omega_1^2 - \omega_0^2)^2(\omega_2^2 - \omega_0^2)^2} \tag{1-16}$$

In the same manner one finds

$$\chi_{xxxx}(\omega_1 = \omega_1 + \omega_2 - \omega_2) = \chi^*_{xxxx}(\omega_2 = \omega_2 + \omega_1 - \omega_1) \tag{1-17}$$

and

$$\chi_{xxxx}(2\omega_1 - \omega_2 = \omega_1 + \omega_1 - \omega_2) = \frac{N_0(e^4/m^3)v^2}{D^2(\omega_1)D^*(\omega_2)D(2\omega_1 - \omega_2)D(\omega_1 - \omega_2)} \tag{1-18}$$

Comparison of Eqs. (1-10), (1-12), (1-16) shows that the following order of magnitude relationship exists between the first order nonlinear polarization and the linear polarization and between nonlinear polarizations of successive orders,

$$|P^{NL}/P^{L}| \approx |P^{NL}_{(n+1)}/P^{NL}_{(n)}| \approx \frac{e|E|}{mD}\frac{v}{D}$$

It may be expected from the physical nature of electronic binding that if the deviation x is of the order of the radius a of the equilibrium orbital of the electron, the nonlinear force mvx^2 is of the same order as the linear force, $m\omega_0^2 a = e|E_{at}|$, where E_{at} is the intra-atomic electric field binding the electron. Therefore,

$v/D \approx v/\omega_0^2 \approx a^{-1}$ and the ratio of the magnitudes of polarization in successive orders is $e|E|/m\omega_0^2 a = |E|/|E_{at}|$. The electric field amplitude of the light wave must be compared with the electric field inside the atom, which is typically of the order of 3×10^8 volts/cm. Therefore, even for the extreme power flux densities of 10^{10} watts/cm^2 in the focus of a Q-switched laser, the nonlinear response can still be treated as a small perturbation, since $|E/E_{at}| \sim 3 \times 10^{-3}$ in this extreme case. It should be noted that the ratio is enhanced by a factor $Q = \omega_0/\Gamma$, whenever a resonance in one of the factors in the denominator occurs. It should also be noted that even if the magnitude of a nonlinear effect is small, it may still be detectable due to the excellent discrimination in optical experiments. Terhune[11] has, for example, detected third-harmonic generation, even though only one out of every 10^{15} photons was converted in this manner. Since a typical ruby laser pulse contains 4×10^{18} photons representing one joule, a thousand ultraviolet photons per pulse at $\lambda = 2313$ A are readily detectable. The polarization at the third-harmonic frequency 3ω, is of course given by a similar susceptibility as (1-15) except that the denominator now contains the factors $D^3(\omega_1)\,D(2\omega_1)\,D(3\omega_1)$.

Magnetic Gyroscopes

Many nonlinear effects have first been discussed in connection with magnetic resonance.[12,13,14,15] This phenomenon can be described in a classical manner as the motion of a magnetic gyroscope without moment of inertia, having a constant ratio γ between its angular momentum and its magnetic moment. The Bloch equations of motion with phenomenological damping terms are,

$$\begin{aligned} \dot{M}_{x,y} &= \gamma(\mathbf{M} \times \mathbf{H})_{x,y} - M_{x,y}/T_2 \\ \dot{M}_z &= \gamma(\mathbf{M} \times \mathbf{H})_z - (M_z - M_0)/T_1 \end{aligned} \qquad (1\text{-}19)$$

Apply a magnetic field with a large dc component H_0 and a small oscillating component at frequency ω_2 in the z-direction and another component precessing at a frequency ω_1 around the z-direction

$$\begin{aligned} H_z &= H_0 + H_2 \exp(-i\omega_2 t) + H_2^* \exp(+i\omega_2 t) \\ H_x + iH_y &= 2H_1 \exp(-i\omega_1 t) \end{aligned} \qquad (1\text{-}20)$$

The magnetic resonant frequency is $\omega_0 = \gamma H_0$.

Again one may solve for the Fourier components for each of the vector components M_x, M_y, and M_z of the magnetic moment. When an expansion in terms of ascending powers of the amplitudes H_1 and H_2 is made, one finds in this manner for the lowest order nonlinear response,

$$M_Z^{NL}(0) = -2\gamma^2 |H_1|^2 T_1 T_2 M_0 / \{1 + (\omega_1 - \omega_0)^2 T_2^2\}^2 \tag{1-21}$$

$$\left(M_x + iM_y\right)^{NL}(\omega_1 - \omega_2) = \frac{-2\gamma^2 H_1 H_2^* M_0 \exp - i(\omega_1 - \omega_2)t}{(\omega_1 - \omega_0 + iT_2^{-1})(\omega_1 - \omega_2 - \omega_0 + iT_2^{-1})} \tag{1-22}$$

Eq. (1-21) represents the onset of the familiar power saturation, and Eq. (1-22) shows the generation of a difference frequency. There is a similar expression for the sum frequency if ω_2 is replaced by $-\omega_2$.

In the next higher nonlinear approximation these Fourier components beat again with the frequencies of the incident fields. These terms will not be reproduced here. They are very similar to the ones derived for the harmonic oscillator in this approximation. Furthermore, they are identical to quantum mechanical expressions derived in the next chapter, when these are applied to a system with only two energy levels.

1-3 PHENOMENOLOGICAL INTERPRETATION OF THE NONLINEAR POLARIZATION

The physical interpretation of these complex nonlinear susceptibilities in terms of elementary quantum processes falls clearly outside the scope of this classical introduction. It is, however, possible to give a phenomenological interpretation. Consider first the lowest order nonlinearity

$$\mathbf{P}^{NL}_{(\omega_3)} = \boldsymbol{\chi}(\omega_3 = \omega_1 + \omega_2)\, \mathbf{E}_1 \mathbf{E}_2 \exp\{i(\mathbf{k}_1 + \mathbf{k}_2)\cdot \mathbf{r} - i(\omega_1 + \omega_2)t\} \tag{1-23}$$

In general $\boldsymbol{\chi}$ is a third-rank tensor. It vanishes for any system with a center of inversion. The nonvanishing elements in crystals which lack inversion symmetry are the same as the nonvanishing elements of the piezoelectric tensor.† The derivation of the form of these tensors for the various point group symmetries has been treated by many authors.[16]

If one of the applied electric fields is a dc field, say $\omega_2 = 0$, Eq. (1-23) describes the linear Kerr effect in a medium that may be absorbing. It is well known that the time-averaged work done per unit

†If $\omega_1 = \omega_2$, the symmetry properties of the tensor $\chi(\omega_3 = \omega_1 + \omega_2)$ are identical to those of the piezoelectric tensor. Both are symmetric in the last two indices. For $\omega_1 \neq \omega_2$, the susceptibility tensor is not necessarily symmetric in these indices.

volume by a harmonic field at frequency ω on a dielectric medium is given by

$$W = \frac{1}{T}\int_0^T \mathbf{E}\cdot\frac{d\mathbf{P}}{dt}\,dt = -2\omega\ \mathrm{Im}\ \mathbf{E}\cdot\mathbf{P}^* = +2\omega\ \mathrm{Im}\ \mathbf{E}^*\cdot\mathbf{P} \qquad (1\text{-}24)$$

if T contains a large number of cycles. For the linear case this becomes simply

$$W = 2\omega\chi''(\omega)\,|E|^2$$

Note that a positive absorption occurs for a positive value of χ'' in $\chi^L = \chi' + i\chi''$ with our convention that a positive frequency has a time-dependence $\exp(-i\omega t)$. For the electro-optic Kerr effect one has an additional power absorption

$$W = 2\omega\chi''(\omega = \omega + 0)\,|E|^2 E_{dc}$$

For the second-harmonic generation the imaginary part of $\chi(2\omega = \omega + \omega)$ does not imply absorption. The complex value of this parametric susceptibility merely relates the phase of the second-harmonic polarization to the phase of the light field at the fundamental frequency. There is no time-averaged absorbed power or time-averaged stored energy of a component of polarization at 2ω in a field at ω. Note that the spatial variation of the phase of the second-harmonic polarization is determined by the wave vector of the fundamental wave and is given by $\exp(2i\mathbf{k}_1\cdot\mathbf{r})$ according to Eq. (1-23).

It is possible to define a thermodynamic potential function F from which the polarization may be derived by differentiation. This function may be identified with the free enthalpy or the capability of the medium to do work at constant entropy on the generators of the field. If the medium is lossless, no heat is developed and the entropy of the medium is constant. It is useful to consider only the time-averaged function, which has a total differential,

$$\langle dF\rangle_t = -\sum_i \mathbf{P}(\omega_i)\cdot d\mathbf{E}_i^* - \sum_i \mathbf{P}^*(-\omega_1)\cdot d\mathbf{E}_i \qquad (1\text{-}25)$$

For a linear medium $\langle F^L\rangle$ takes on the well-known form quadratic in the field strengths,

$$\langle F^L\rangle = -\mathrm{Re}\sum_i \mathbf{E}^*(-\omega_i)\cdot\chi^L(\omega_i)\cdot\mathbf{E}(\omega_i) \qquad (1\text{-}26)$$

The lowest order nonlinearity corresponds to a cubic term in the Fourier amplitudes. Consider, for example, the case of three waves at frequencies ω_1, ω_2, and $\omega_3 = \omega_1 + \omega_2$. The time-averaged potential will have a term

$$\langle F^{NL} \rangle = -2 \,\mathrm{Re}\, \{\mathbf{E}_3^*(\omega_3)\, \chi^{NL}(\omega_3 = \omega_1 + \omega_2)\mathbf{E}_1(\omega_1)\mathbf{E}_2(\omega_2)\} \tag{1-27}$$

One finds in this manner the ith component of the polarization at ω_3 due to the jth component of the field at ω_1 and the kth component of the field at ω_2.

$$P_i(\omega_3) = -\partial F/\partial E_{i3}^* = \chi_{ijk}(\omega_3 = \omega_1 + \omega_2)E_{j1}E_{k2}$$

One also finds, since χ is real for the lossless medium,

$$P_j(\omega_1) = -(\partial F/\partial E_{j1})^* = \chi_{ijk}(\omega_3 = \omega_1 + \omega_2)E_{i3}E_{k2}^*$$

This leads to a permutation symmetry relation

$$\begin{aligned} \chi_{ijk}(\omega_3 = \omega_1 + \omega_2) &= \chi_{jik}(\omega_1 = \omega_3 - \omega_2) \\ &= \chi_{kij}(\omega_2 = \omega_3 - \omega_1) \end{aligned} \tag{1-28}$$

For real nonlinear susceptibilities one may interchange the tensor indices, provided the corresponding frequencies are also interchanged. If the dispersion in the whole frequency range containing ω_1, ω_2, and ω_3 is negligible, the frequencies may be interchanged at random. In this case the nonlinear tensor becomes symmetric in all indices which are freely interchangeable. These relations for the dispersionless case were first formulated by Kleinman,[17] whereas the general permutation symmetry relations (1-28) were given by Armstrong, Bloembergen, Ducuing, and Pershan.[18] They may be verified by explicit calculation in each particular case. They follow, e.g., for the anharmonic oscillator with negligible damping.

The component of the nonlinear polarization at frequency ω_i which is 90° out of the phase with the field at ω_i is responsible for a net power loss or gain at this frequency. The time-averaged work done by the wave E_3 at the frequency ω_3 is, for example, given by

$$W(\omega_3) = -i\omega_3(\mathbf{E}_3 \cdot \mathbf{P}_3^* - \mathbf{E}_3^* \cdot \mathbf{P}_3) = 2\omega_3 \,\mathrm{Im}\, \mathbf{E}_3^* \cdot \mathbf{P}_3$$

The total work done by all three waves on the medium which represents the rate of heat development is given by

$$\begin{aligned} W = -2\mathrm{Im} \sum_{ijk} [&\omega_3 \chi_{ijk}^*(\omega_3 = \omega_1 + \omega_2) - \omega_1 \chi_{jik}(\omega_1 = \omega_3 - \omega_2) \\ &- \omega_2 \chi_{kij}(\omega_2 = \omega_3 - \omega_1)] E_{1j}^* E_{2k}^* E_{3i} \end{aligned} \tag{1-29}$$

For a lossless medium with real χ satisfying the symmetry relations

(1-28) this expression properly vanishes. Any power that is absorbed at ω_3 is emitted at ω_1 and ω_2, and vice versa, in this approximation. This corresponds to parametric conversion of power between the frequencies ω_1, ω_2, and ω_3, described by the cubic term.

If initially only waves ω_1 and ω_2 are present, they generate a phased array of dipoles at the frequency $\omega_3 = \omega_1 + \omega_2$. The radiation field ω_3 from this array will be calculated in a later chapter. In this first part we consider the fields as given and calculate the induced polarizations, linear and nonlinear. Later, these polarizations must be in turn considered as sources of the fields, and self-consistent solutions must be found with the aid of Maxwell's equations.

Consider the case of second-harmonic generation by a polarization $P^{NL}(2\omega) = \chi(2\omega = \omega + \omega) E_1^2 \exp(2i\mathbf{k}_1 \cdot \mathbf{r} - 2i\omega t)$ in the presence of a wave $E_2 \exp(i\mathbf{k}_2 \cdot \mathbf{r} - 2i\omega t)$. Due to the natural color dispersion the wave at 2ω will travel in general with a different propagation velocity as the nonlinear source since $\mathbf{k}_2 \neq 2\mathbf{k}_1$. The wave and the source will get out of step after a distance $\mathbf{l}_{coh}$ for which $(\mathbf{k}_2 - 2\mathbf{k}_1) \cdot \mathbf{l}_{coh} = \pi$. Therefore, the power absorption from the second harmonic wave $-W(\omega) = W(2\omega) = 2\omega \operatorname{Im} E_2^* P^{NL}(2\omega)$ will alternate in sign. After the second harmonic has been generated with a corresponding decrease in power of the fundamental wave, the harmonic will be reabsorbed and the fundamental recreated, etc. This interference effect is vividly demonstrated as the optical thickness of the nonlinear crystal in the arrangement of Figure 1-1 is changed by rotation or by variation of the temperature. A characteristic pattern is shown in Figure 1-2, obtained by Terhune and coworkers.[19] The distance between two consecutive maxima corresponds to an increase in optical path by $2l_{coh}$. Giordmaine[20] and Terhune[19] have shown how this mismatch of the phase velocity due to color dispersion may be compensated by the anisotropy of the phase velocity of the extraordinary ray in uniaxial crystals. Figure 1-3 demonstrates the enormous increase in second-harmonic intensity, when the phase velocity of the second-harmonic extraordinary ray is matched to the phase velocity of ordinary fundamental ray.

If ω_1 is a light frequency and ω_2 a microwave frequency, the polarization at the frequencies $\omega_1 \pm \omega_2$ generates sidebands on the light carrier. This microwave modulation of light beams is important in communications applications.[21] Phase matching is again of importance if the structure is larger than one wave length at the modulation frequency. The inverse, in which two light beams beat to generate microwave power at the difference frequency, has also been observed.

The sum frequency of two different light beams is also observable. Finally Eq. (1-23) also describes the rectification of light if one takes $\omega_2 = -\omega_1$ and $E_2 = E_1^*$. A dc voltage is developed across the crystal

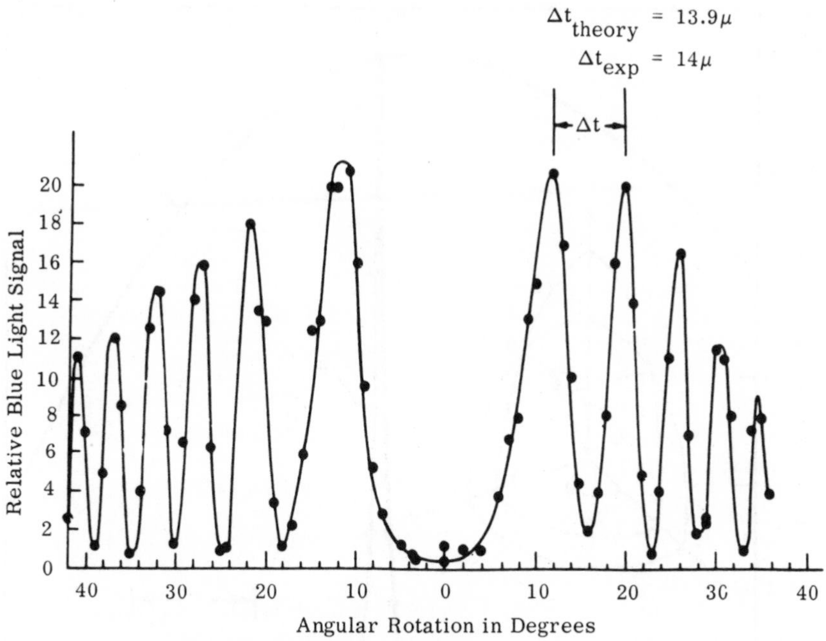

Figure 1-2. The second harmonic intensity is a periodic function of the optical thickness of the piezoelectric crystal, when there is a mismatch in the phase velocities of the fundamental and the harmonic wave (after Maker et al.[19]).

as observed by Franken.[22] It may be considered as the inverse linear electro-optic effect.

Turning to the next higher order nonlinearity

$$\mathbf{P}^{NL}(\omega_1 + \omega_2 + \omega_3)$$
$$= \chi \mathbf{E}_1 \mathbf{E}_2 \mathbf{E}_3 \exp\{i(\mathbf{k}_1 + \mathbf{k}_2 + \mathbf{k}_3)\cdot\mathbf{r} - i(\omega_1 + \omega_2 + \omega_3)t\}$$

one notes immediately that the fourth-rank tensor does not vanish in any point group symmetry. A variety of effects described by this nonlinearity exists also in isotropic fluids.

If one takes $\omega_1 = \omega_2 = \omega_3$, one obtains third-harmonic generation, first observed by Terhune. For $\omega_1 = \omega_2 = -\omega_3$, $\mathbf{E}_1 = \mathbf{E}_2 = \mathbf{E}_3^*$ one finds a term in the polarization

$$P_i^{NL}(\omega_1) = \sum \chi_{ijk\ell}(\omega_1 = -\omega_1 + \omega_1 - \omega_1) \times E_j^*(-\omega_1) E_k(\omega_1) E_\ell(+\omega_1) \tag{1-30}$$

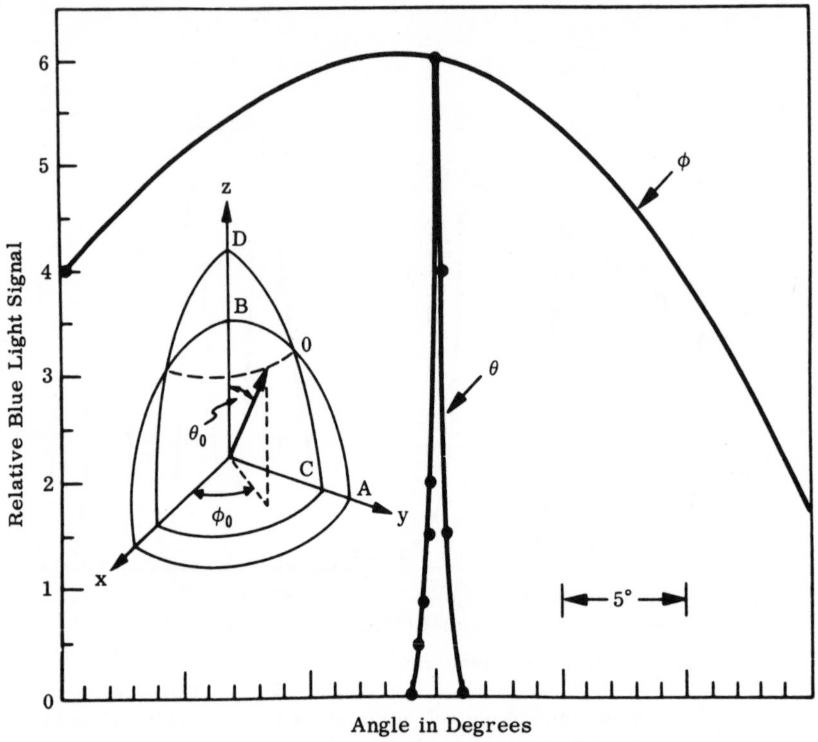

Figure 1-3. Increased production of second harmonic intensity in a crystal of KH_2PO_4, when the phase velocity of the extraordinary ray at 2ω is matched to the ordinary ray at ω (after Maker et al.[19]).

The real part of this susceptibility therefore describes a change in the index of refraction proportional to the intensity of the light beam. In the case of a lossy medium, the imaginary part will describe a change in the absorption coefficient with intensity. One part of this represents the beginning of a saturation effect, well known in magnetic resonance. It is used in saturable filters, which become transparent for high-power fluxes.[23] It is large when ω_1 is near a resonance of the system. The other part is large when $2\omega_1$ is near resonance. This term represents two quanta absorption. It has been detected by Garrett[24] and others. A green fluorescence of Eu^{2+} ions in CaF_2 was detected by irradiation with ruby light. The red quanta could not excite the ion in separate successive steps. The observed fluorescence was indeed proportional to the square of the intensity.

In an isotropic fluid the fourth-rank tensor has only two independent elements. Eq. (1-30) can in this case be written in the vector form, using the Voigt abbreviations 1 = xx, 2 = yy, 4 = xy

$$\mathbf{P}^{NL}(\omega) = \chi_{44}(\omega)\,\mathbf{E}(\omega)\,(\mathbf{E}(\omega)\cdot\mathbf{E}^*(-\omega)) + \chi_{12}\,\mathbf{E}^*(-\omega)(\mathbf{E}(\omega)\cdot\mathbf{E}(\omega)) \tag{1-31}$$

Terhune has demonstrated the existence of the χ_{12} term. This experiment will be discussed more fully in Chapter 5.

If one takes $\omega_1 > \omega_2 = -\omega_3$ and assumes for simplicity that E_2 is linearly polarized in the x-direction, the nonlinear polarization at frequency ω_1 has the following components:

$$P_x^{NL}(\omega_1) = \chi_{11}E_{1x}|E_{2x}|^2$$

$$P_y^{NL}(\omega_1) = \chi_{21}E_{1y}|E_{2x}|^2$$

$$P_z^{NL}(\omega_1) = \chi_{21}E_{1z}|E_{2x}|^2$$

The real part of these susceptibility components represents a change of index of refraction at frequency ω_1 in the presence of an intensity at ω_2. If $\omega_2 = 0$, the term just describes the quadratic Kerr effect with an induced birefringence.

The imaginary part of χ represents a positive absorption by the material for $\chi'' > 0$, proportional to the intensity at ω_2. Note that the phase of this nonlinear polarization is fixed with respect to the phase of the field at the same frequency.

The change in the time-averaged free energy density, due to the simultaneous presence of the fields at ω_1 and ω_2 and biquadratic in the field amplitudes can be written as

$$F^{NL} = -2\,\mathrm{Re}\,\chi_{11}|E_{1x}|^2|E_{2x}|^2 - 2\,\mathrm{Re}\,\chi_{12}|E_{2x}|^2(|E_{1y}|^2 + |E_{1z}|^2)$$

The time-averaged net absorbed power by the material is expressed by

$$W = 2(\omega_1 - \omega_2)\{\mathrm{Im}\,\chi_{11}(\omega_1 = \omega_1 + \omega_2 - \omega_2)|E_{1x}|^2|E_{2x}|^2 + \mathrm{Im}\,\chi_{12}(|E_{1y}|^2 + |E_{1z}|^2)|E_{2x}|^2\}$$

Use has been made of a symmetry relation for the complex susceptibilities of the type given by Eq. (1-17). This relationship holds if only one resonant frequency near $\omega_2 - \omega_1$ is of importance and the damping term Γ need only be retained in one of the denominators. This condition is satisfied in the anharmonic oscillator model.

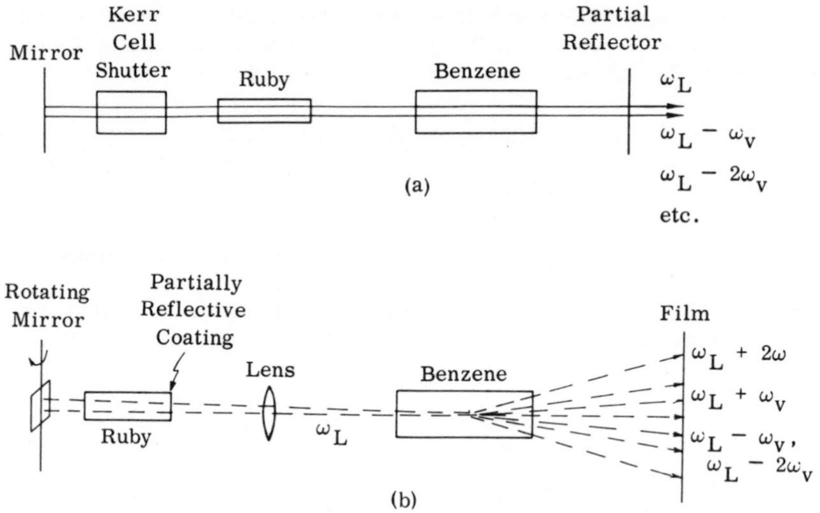

Figure 1-4. (a) Experimental arrangement of a Raman laser.
(b) Experimental arrangement for the production of stimulated stokes and antistokes light in an externally focused laser beam.

Since $\text{Im}\, \chi_{11}(\omega_2 = \omega_2 + \omega_1 - \omega_1) < 0$ according to Eq. (1-17), there is a negative absorption at the lower frequency ω_2 and a positive absorption at the higher frequency ω_1. There is an exponential gain factor at ω_2, proportional to the intensity at ω_1. If $|E(\omega_1)|^2$ is taken large enough, this gain may overcome all losses at ω_2. If the gain is sufficiently large and some feedback is provided, an oscillator at ω_2 will result. Although the classical picture gives $P^{NL}(\omega_2) = 0$ for $E(\omega_2) = 0$, a small disturbance or noise would put the system into oscillation. A proper description of the oscillation build-up would be possible only with quantum mechanics, since spontaneous emission must be invoked, which might be considered in a nonrigorous fashion as noise from zero point vibrations. This is amplified and builds up the oscillation.

The effect described corresponds to the Raman effect, where a quantum $\hbar\omega_1$ is absorbed, $\hbar\omega_2$ is emitted and the energy of the system is raised by $\hbar\omega_0 = \hbar(\omega_1 - \omega_2)$. The Raman laser effect was discovered by Woodbury et al.[25] They found that a ruby laser with a liquid cell between the resonator plates emits radiation at a frequency which differs from the ruby laser frequency by the frequency of a vibrational resonance of the molecule if the laser is operated above a certain power level.

The effect is quite pronounced once the threshold is exceeded. It has been found in many liquids, as well as in solids and in gases. It can be obtained also in an external laser beam, which may be focused into a crystal or into a cell containing the fluid under investigation. If both fields at the laser frequency ω_1 and the Stokes frequency ω_2 are present, there will also be a polarization at the antistokes frequency $2\omega_1 - \omega_2$. This term is given by the susceptibility Eq. (1-18) and makes use of the same resonant denominator. Terhune[26] was the first to detect the radiation at the antistokes frequency. It is emitted in a characteristic ring pattern. There will be clearly polarizations at other combination frequencies $\omega_1 + n(\omega_1 - \omega_2)$, $\omega_1 - n(\omega_1 - \omega_2)$, etc. These Raman-type effects will be discussed in more detail in Chapter 5. A typical experimental arrangement is shown in Figure 1-4.

1-4 SYNOPSIS

In this chapter the nonlinear polarization has been introduced and expressed in terms of the applied fields, which are assumed to be prescribed. The next logical step is to consider the nonlinear polarization as additional sources for the fields. This second step will be taken in Chapter 3. The reader who does not wish to interrupt the classical exposé may skip the next chapter which gives a quantum mechanical derivation of the nonlinearity susceptibility to gain some further insight into the atomic mechanisms of the nonlinearity. At the same time the restriction to the pure electric dipole case will be lifted. Furthermore, some general relationships between real and imaginary parts of the susceptibilities and fundamental quantum processes will be pointed out and the limitations of the semiclassical treatments of the fields are briefly discussed.

In Chapter 3 the nonlinearities are incorporated in Maxwell's equations. The interaction between light waves in a nonlinear medium is given in Chapter 4. Solution of the coupled wave equations are given for parametric and Raman-type effects. The considerations are extended to the coupling of light and vibrational waves. Generalizations of the optical laws of reflection, refraction, etc. follow from the analysis of the boundary conditions for a nonlinear medium.

Experimental results and their interpretation are briefly reviewed in Chapter 5. Nonlinear susceptibility data for various materials are discussed. Particular attention is also given to an analysis of the data on the Raman effect. Some special topics, including nonlinearities in laser oscillators, are mentioned in Chapter 6. The inclusion of reprints of three papers as three appendices made it possible to omit certain rather complex calculations from the main text. Since the supply of reprints of the 1962 papers has been exhausted, it was thought worthwhile to make them again available in this form.

REFERENCES

1. P. A. Franken, A. E. Hill, C. W. Peters, and G. Weinreich, *Phys. Rev. Letters*, **7**, 118 (1961).
2. P. A. Franken and J. F. Ward, *Rev. Mod. Phys.*, **35**, 23 (1963).
3. N. Bloembergen, *Proc. IEEE*, **51**, 124 (1963).
4. P. S. Pershan, *Progress in Optics*, ed. E. Wolf, North-Holland Publishing Co., Amsterdam, to be published. The author is indebted to Professor Pershan for making his manuscript available before publication.
5. F. J. McClung and R. W. Hellwarth, *Proc. IEEE*, **51**, 46 (1963).
6. J. F. Ward and P. A. Franken, *Phys. Rev.*, **133**, A 183 (1964).
7. B. Lax, J. G. Mavroides, and D. F. Edwards, *Phys. Rev. Letters*, **8**, 166 (1962).
8. R. F. Whitmer and E. B. Barrett, *Phys. Rev.*, **121**, 661 (1961); R. F. Whitmer and E. B. Barrett, *Phys. Rev.*, **125**, 1478 (1962).
9. R. Becker, *Elektronen Theorie*, Teubner, Leipzig, 1933.
10. Lord Rayleigh, *Theory of Sound*, **I**, 76 ff, Dover Publications, New York, 1945.
11. R. W. Terhune, P. D. Maker, and C. M. Savage, *Phys. Rev. Letters*, **8**, 404 (1962).
12. J. Brossel, B. Cagnac, and A. Kastler, *J. Phys. Rad.*, **15**, 6 (1954).
13. P. Kusch, *Phys. Rev.*, **93**, 1022 (1954).
14. F. Bloch, *Phys. Rev.*, **102**, 104 (1956), and many other references quoted in Appendix 3 of this volume.
15. W. P. Ayres, P. H. Vartanian, and J. L. Melchor, *J. App. Phys.*, **27**, 188 (1956).
16. See, for example, J. F. Nye, *Physical Properties of Crystals; their representation by tensors and matrices*, Clarendon Press, Oxford, 1957; or C. S. Smith in *Solid State Physics*, Vol. **5**, ed. F. Seitz and D. Turnbull, Academic Press, New York, 1958, p. 175.
17. D. A. Kleinman, *Phys. Rev.*, **126**, 1977 (1962).
18. J. A. Armstrong, N. Bloembergen, J. Ducuing, and P. S. Pershan, *Phys. Rev.*, **127**, 1918 (1962). This paper is reproduced at the end of this volume and will henceforth be referred to as Appendix I. Compare also, V. I. Karpman, *J. Exp. Theor. Phys. (U.S.S.R.)*, **44**, 1307 (1963), English translation *JETP*, **17**, 882 (1963).
19. P. D. Maker, R. W. Terhune, M. Nisenhoff, and C. M. Savage, *Phys. Rev. Letters*, **8**, 21 (1962).
20. J. A. Giordmaine, *Phys. Rev. Letters*, **8**, 19 (1962).
21. I. P. Kaminow and J. Liu, *Proc. IEEE*, **51**, 132 (1963) and C. J. Peters, *Proc. IEEE*, **51**, 147 (1963).

22. M. Bass, P. A. Franken, J. F. Ward, and G. Weinreich, *Phys. Rev. Letters*, **9**. 446 (1962).
23. P. P. Sorokin, J. J. Luzzi, J. R. Lankard, and G. D. Pettit, *IBM Journal Res.*, **8**, 182 (1964).
24. W. Kaiser and G. C. B. Garrett, *Phys. Rev. Letters*, **7**, 229 (1961).
25. G. Eckhardt, R. W. Hellwarth, F. J. McClung, S. E. Schwarz, D. Weiner, and E. J. Woodbury, *Phys. Rev. Letters*, **9**, 455 (1962).
26. R. W. Terhune, *Bull. Am. Phys. Soc.*, **II** (8), 359 (1963); *Solid State Design*, **4**, 38 (November 1963).

2

QUANTUM THEORY OF NONLINEAR SUSCEPTIBILITIES

2-1 THE LIOUVILLE EQUATION FOR THE DENSITY MATRIX[1]

A quantum mechanical system described by a Hamiltonian $\mathcal{H}_0$ has time-independent eigenstates u_n with energy W_n which are solutions of the Schrödinger equation

$$\mathcal{H}_0 u_n = W_n u_n \tag{2-1}$$

The u_n form a complete set of orthonormal functions. The response of this system to a time-dependent perturbation $\mathcal{H}_1(t)$ is described by a time-dependent wave function, which can be written as a series expansion in the u_n with time-dependent coefficients,

$$\psi(t) = \sum_n c_n(t)\, u_n \tag{2-2}$$

The coefficients c obey the differential equations,

$$\dot{c}_k = -ih^{-1} \sum_n c_n (k \,|\, \mathcal{H}_0 + \mathcal{H}_1(t) \,|\, n) \tag{2-3}$$

The expectation value of an operator O at time t is given by

$$\langle O(t) \rangle = (\psi^*(t)\,|O|\,\psi(t)) = \sum_{k,m} O_{mk}\, c_k\, c_m^* = \operatorname{Tr} O R \tag{2-4}$$

where R is an operator with matrix elements $R_{km} = c_k c_m^*$. It follows readily from Eq. (2-3) that R obeys the differential equation,

$$\dot{R} = i\hbar^{-1}(R\mathcal{H} - \mathcal{H}R) = -i\hbar^{-1}[\mathcal{H}, R] \tag{2-5}$$

This operator form is valid in any system of representation and is especially useful when an ensemble of systems must be described all subjected to the identical Hamiltonian $\mathcal{H} = \mathcal{H}_0 + \mathcal{H}_1(t)$. The ensemble average of $\overline{R} = \rho$ is called the density matrix, which satisfies the equation of motion,

$$\dot{\rho} = -i\hbar^{-1}[\mathcal{H}, \rho] \tag{2-6}$$

If the average initial conditions are known for the ensemble, solution of this equation will give the ensemble expectation value of any physical quantity associated with an operator O

$$\langle O \rangle = \mathrm{Tr}\,(\rho O) \tag{2-7}$$

Since $\mathcal{H}_0$ is independent of time, and $\mathcal{H}_1(t)$ may be considered as a small perturbation, it is possible to obtain a solution in succesive powers of $\mathcal{H}_1$. Transform to the interaction representation

$$\begin{aligned} \rho' &= e^{+(i/\hbar)\mathcal{H}_0 t}\, \rho\, e^{-(i/\hbar)\mathcal{H}_0 t} \\ \mathcal{H}_1' &= e^{+(i/\hbar)\mathcal{H}_0 t}\, \mathcal{H}_1\, e^{-(i/\hbar)\mathcal{H}_0 t} \end{aligned} \tag{2-8}$$

Then ρ' obeys the equation

$$d\rho'/dt = -i\hbar^{-1}(\mathcal{H}_1'\rho' - \rho'\mathcal{H}_1') \tag{2-9}$$

which can be solved by successive approximations,

$$\begin{aligned} \rho'(t) = \rho'(o) - i\hbar^{-1}\int_0^t [\mathcal{H}_1'(t'), \rho'(o)]\,dt' - \hbar^{-2}\int_0^t dt' \int_0^t dt'' \\ \times [\mathcal{H}_1'(t'), [\mathcal{H}_1'(t''), \rho'(o)]] + \cdots \end{aligned} \tag{2-10}$$

One may, of course, transform back to the Heisenberg representation $\rho(t)$. This concludes the formal solution of time-dependent perturbation theory.

2-2 RANDOM PERTURBATIONS AND DAMPING

As shown in Chapter 1 our main interest will be in the steady state response to perturbations by periodic electromagnetic fields. Any system that will be considered is, however, subjected to unavoidable perturbations of a stochastic nature. On the average these perturbations are responsible for the damping terms that were introduced in a phenomenological fasion in the classical equations of motion. The physical origin of these random perturbations is diverse. Thermal agitation in fluids, lattice vibrations in crystals, spontaneous emission of light, radiationless decay by spontaneous emission of phonons, collisions with conduction electrons, ionic or molecular collisions in a gas all fall into this category. In the semi-classical treatment the random perturbation $\mathcal{H}_1(t)$ is an operator which only acts on the material system under consideration. The motion of the electromagnetic fields, vibrations, motions of other particles, is described classically in a stochastic sense. The average $\langle \mathcal{H}_1'(t)\rangle_{Av} = 0$, i.e., all matrix elements $\langle (m|\mathcal{H}_1'|n)\rangle_{Av}$ vanish on the average. If there is any nonvanishing part, it may be included in $\mathcal{H}_0$. The stochastic process is assumed to be stationary, so that correlation functions such as $\langle \mathcal{H}_{1,k\ell}(t)\mathcal{H}_{1,mn}(t+\tau)\rangle_{Av}$ are independent of t. Spectral densities can be defined as the Fourier transforms of these correlation functions. The essential point is that these spectral distributions of the stochastic process are sufficiently broad so that a short correlation time τ_c can be defined, such that

$$\overline{\mathcal{H}_{1,k\ell}'(t)\mathcal{H}_{1,mn}'(t+\tau)} = 0 \quad \text{for} \quad \tau > \tau_c$$

Consider the solution, Eq. (2-10), of Eq. (2-9) for times $t \gg \tau_c$. Take the average over all damping histories. It is permissible to ignore the correlation between $\rho'(o)$ and $\mathcal{H}(t')$. In principle the initial value $\rho'(o)$ is determined by immediately preceding values of $\mathcal{H}(t)$ and these are in turn correlated with immediately following values $\mathcal{H}(t')$. The correlation between $\mathcal{H}(t')$ and $\rho'(o)$ vanishes, however, for $t' > \tau_c$ and most of the contributions to the integrals comes from values for which $t'' > \tau_c$, $t' > \tau_c$, but $t' - t'' < \tau_c$. Transform from the variable t'' to $\tau = t' - t''$. The limit of integration over τ may be extended to $+\infty$ for the same reasons. If higher order terms in Eq. (2-10) are negligible, the average solution for a random perturbation becomes, for times $t \gg \tau_c$,

$$\rho'(t) = \rho'(o) - t\hbar^{-2}\int_0^\infty \overline{[\mathcal{H}_1'(t), [\mathcal{H}_1'(t-\tau), \rho'(o)]]}\, d\tau \tag{2-11}$$

The relative change in $\rho'(t)$ is given by

$$\frac{\rho'(t) - \rho'(o)}{\rho'(o)} \approx \frac{t}{\rho'(o)}\frac{d\rho'}{dt} \approx -t\hbar^{-2}\,\overline{|\mathcal{H}_1|^2}\,\tau_c \tag{2-12}$$

If the mean square perturbation is sufficiently weak and rapidly varying such that

$$\hbar^{-2}|\mathcal{H}_1|^2\tau_c^2 \ll 1 \tag{2-13}$$

one may choose $t \gg \tau_c$ and yet satisfy the relation $\left|\frac{\rho'(t) - \rho'(o)}{\rho'(o)}\right| \ll 1$.

Higher order terms are indeed negligible under these same conditions, because the ratio of the magnitudes between two successive approximations is apparently $(\hbar^{-2}\langle|\mathcal{H}_1|\rangle^2_{Av}\tau_c^2)^{1/2}$. The condition (2-13) is therefore sufficient to rewrite (2-11) in the form

$$\left(\frac{d}{dt}\right)_{random} \rho' = -\hbar^{-2}\int_0^\infty \overline{[\mathcal{H}_1'(t), [\mathcal{H}_1'(t-\tau), \rho'(t)]]}\, d\tau \qquad (2\text{-}14)$$

If this operator equation is written out explicitly in the representation for which $\mathcal{H}_0$ is diagonal, one finds for the rate of change of a typical density matrix element,

$$\frac{d}{dt}\rho'_{k\ell} = \sum_{mn} e^{i(\omega_{km}-\omega_{\ell n})t} R_{k\ell,mn}\rho'_{mn}$$

where the stationarity of Eq. (2-14) ensures that the coefficients R are independent of t. If very small secular perturbations are ignored, i.e., only nonperiodic terms are kept for which $\omega_{km} = \omega_{\ell n}$, one obtains a characteristic time-independent damping matrix

$$\frac{d}{dt}\rho'_{k\ell} = \sum_{mn} R_{k\ell,mn}\rho'_{mn} \qquad (2\text{-}15)$$

Barring accidental degeneracies and situations with equal spacings or nearly equal spacings between levels, the only nonsecular terms caused by the stochastic perturbation are the damping terms,

$$\frac{d}{dt}\rho'_{k\ell} = R_{k\ell k\ell}\rho'_{k\ell}$$

$$\frac{d}{dt}\rho'_{kk} = \sum_m R_{kkmm}\rho'_{mm} = \sum_m w_{km}\rho'_{mm} - \left(\sum w_{mk}\right)\rho'_{kk} \qquad (2\text{-}16)$$

The last equation corresponds to a rate equation for the populations in the different energy levels. The coefficient w_{km} may be identified with a time proportional transition probability between levels k and m. By working out the commutators in Eq. (2-14), one finds†

†In Eqs. (2-15) and (2-17) the damping constants are expressed in terms of the perturbation in the Schrödinger representation, with exponential time factors explicitly exhibited. The density matrix itself is still considered in the interaction representation.

$$w_{km} = \hbar^{-2} \int_{-\infty}^{+\infty} \overline{\mathcal{H}_{km}(t)\mathcal{H}_{mk}(t - \tau)}\, e^{-i\omega_{km}\tau}\, d\tau \tag{2-17}$$

The transition probability is proportional to the spectral density of the connecting matrix element at the transition frequency ω_{km}. All diagonal terms or populations relax in a coupled fashion, leading to $\mathfrak{N} - 1$ characteristic relaxation times for a system with $\mathfrak{N}$ energy levels, satisfying the trace relation $\Sigma \rho_{kk} = 1$.

The damping of the off-diagonal matrix elements $\rho_{k\ell}$ can be obtained in a similar manner by collecting terms in the commutators. Off-diagonal elements in the perturbation make a contribution,

$$R_{k\ell k\ell}^{\text{non-ad}} = -\frac{1}{2} \sum_n (w_{kn} + w_{\ell n}) = R_{\ell k \ell k}^{\text{non-ad}} \tag{2-18}$$

The adiabatic part of the off-diagonal damping is associated with random perturbations of the energy levels by diagonal elements of the stochastic process. In a certain sense, the perturbations modulate the resonant frequency and broaden the resonance in this manner,

$$\begin{aligned} R_{k\ell k\ell}^{\text{ad}} &= -\hbar^{-2} \int_0^\infty \{\overline{\mathcal{H}_{kk}(t)\mathcal{H}_{kk}(t - \tau)} + \overline{\mathcal{H}_{\ell\ell}(t)\mathcal{H}_{\ell\ell}(t - \tau)} \\ &\quad - \overline{\mathcal{H}_{kk}(t)\mathcal{H}_{\ell\ell}(t - \tau)} - \overline{\mathcal{H}_{\ell\ell}(t)\mathcal{H}_{kk}(t - \tau)}\}\, d\tau \\ &= \int_0^\infty \overline{\delta\omega_{k\ell}(t)\, \delta\omega_{k\ell}(t - \tau)\, d\tau} \quad \text{with} \\ \hbar\delta\omega_{k\ell}(t) &= (k|\mathcal{H}_1(t)|k) - (\ell|\mathcal{H}_1(t)|\ell) \end{aligned} \tag{2-19}$$

A most serious shortcoming of this semiclassical theory of relaxation is that the system relaxes according to Eqs. (2-16) to a final state with all diagonal elements equal to each other. This corresponds to thermodynamic equilibrium at infinite temperature. On general physical grounds one should expect the final state to be one in which the system attains the same temperature as the reservoir which provides the random motion of the damping mechanism. Actually a proper quantization of the reservoir will remedy this situation. In Section 2-8 this will be shown explicitly, when the damping produced by spontaneous emission will be treated. For the time being we assume on the basis of the principle of detailed balancing in an *ad hoc* fashion

$$w_{kn} = w_{nk} \exp(-\hbar\omega_{kn}/kT) \tag{2-20}$$

This ensures a Boltzmann distribution for the equilibrium density matrix populations.

$$\rho_{kk} = \rho_{nn} \exp(-\hbar \omega_{kn}/kT) \tag{2-21}$$

The off-diagonal elements decay to zero, corresponding to the condition of random phases of the wave functions in thermodynamic equilibrium. Also note that the expectation values of ρ_{kn} remain zero whenever a stochastic perturbation is switched on. No phase relationships are established on the average by random processes. It should be noted that different damping mechanisms need not be described by the same temperature. A striking example is the case of the pumping of a laser by incoherent radiation. In the solid angle subtended by the flashlamp the incoherent radiation processes have a high temperature, in other directions a much lower one, whereas a radiationless transition involving the lattice is characterized by the temperature of the cyrstal. In a gaseous laser the electron collision processes are described by one temperature, wall collisions by another, and so on. All stochastic processes, including incoherent pumping processes, will be taken into account by the appropriate phenomenological damping terms, when they satisfy the basic condition, Eq. (2-13).

A coherent time-dependent perturbation consists of one or more periodic components which have a very long or infinite correlation time τ_c. The spectral density of this perturbation consists of δ-functions. The transition case with $\hbar^{-1}|\mathcal{H}_1|\tau_c \approx 1$ will not be treated.

When the transformation from the interaction representation back to the Schrödinger representation is performed, the equation of motion becomes

$$\dot{\rho} = -i\hbar^{-1}[\mathcal{H}_0, \rho| - i\hbar^{-1}[\mathcal{H}_{coh}(t), \rho] + R\rho \tag{2-22}$$

where R represents the time-independent damping constants, defined by Eqs. (2-16) and (2-19) for the representation in which $\mathcal{H}_0$ is diagonal. Only under certain special circumstances is it possible to write the damping term in operator form. If a set of energy levels n is considered, which undergo only nonadiabatic perturbations, and if the stochastic perturbation only connects to and from levels μ outside the set, one may write the damping term as $-\frac{1}{2}(\Gamma\rho + \rho\Gamma)$, where Γ is a diagonal matrix and $\Gamma_{nn} = \Sigma_\mu W_{n\mu}$.

In addition to the homogeneous damping discussed here, one often encounters an inhomogeneous broadening caused by a static distribution of resonant frequencies. Strains in crystals may, for example, cause such a broadening. In some cases one gets a reasonable approximation by simply increasing the damping constant $R_{k\ell k\ell}$ to correspond to the observed broadening of the line. This may, however, lead to incorrect results. In principle, inhomogeneous broadening should be taken into account by an integration over the distribution of resonant frequencies at the end of the calculation.

2-3 RESPONSE TO PERIODIC PERTURBATIONS

Whereas the response to the random perturbations has been solved in the time domain, it is algebraically simpler to solve for the steady state response to periodic perturbations in the frequency domain. Again we use a method of successive approximation to obtain a solution in ascending powers of the coherent perturbation.[2]

The equations for the steady state response become algebraic equations. In the first approximation the equations for a typical diagonal and off-diagonal element are

$$-i\omega\rho_{kk}^{(1)} = \sum_n w_{kn}\rho_{nn}^{(1)} - \sum_n w_{nk}\,\rho_{kk}^{(1)} - i\hbar^{-1}\,[\mathcal{H}_{coh}(\omega),\rho^{(0)}]_{kk}$$

$$-i\omega\rho_{k\ell}^{(1)} = -i\,\omega_{k\ell}\rho_{k\ell}^{(1)} - \Gamma_{k\ell}\rho_{k\ell}^{(1)} - i\hbar^{-1}[\mathcal{H}_{coh}(\omega),\rho^{(0)}]_{k\ell} \qquad (2\text{-}23)$$

where $\rho^{(0)}$ is the steady state density matrix solution in the absence of the coherent perturbation. The diagonal elements are not necessarily equal to a Boltzmann distribution. There may be incoherent pumps, and the steady state populations may be altered by them to the extent that some population inversion occurs. Normally the $\rho^{(0)}$ will be given by the distribution corresponding to thermodynamic equilibrium. The off-diagonal elements of $\rho^{(0)}$ always vanish. There is a random distribution of the phases in thermodynamic equilibrium, which remains random in the presence of random pumping.

The first approximation (Eq. (2-23)) gives the linear response, which contains the same Fourier components as are present in the perturbation $\mathcal{H}_{coh}(\omega)$. A second approximation is found by reinserting the linear approximation into the equation of motion. The term $[\mathcal{H}(\omega),\rho^{(1)}]_{kk}$ will, in general, contain combination frequencies of the original perturbation. One finds, for example, components at the second-harmonic frequency from the equations,

$$-2i\omega\rho_{kk}^{(2)}(2\omega) = \sum w_{kn}\rho_{nn}^{(2)} - \sum_n w_{nk}\rho_{kk}^{(1)} - i\hbar^{-1}[\mathcal{H}_{coh}(\omega),\,\rho^{(1)}(\omega)]_{kk}$$

$$-2i\omega\rho_{k\ell}^{(2)}(2\omega) = -i\omega_{k\ell}\rho^{(2)}(2\omega) - \Gamma_{k\ell}\rho_{k\ell}^{(2)} - i\hbar^{-1}[\mathcal{H}_{coh}(\omega),\,\rho^{(1)}(\omega)]_{k\ell} \qquad (2\text{-}24)$$

This process may be repeated indefinitely. Quite generally a Fourier component at the frequency $\omega_n = \omega_{n-1} + \omega$ in the nth approximation, proportional to the nth power of the perturbation, may be found from the Fourier components at ω_{n-1} in the next lower approximation, e.g.,

$$-i\omega_n \rho_{k\ell}^{(n)}(\omega_n) = -i\omega_{k\ell}\rho_{k\ell}^{(n)}(\omega_n) - \Gamma_{k\ell}\rho_{k\ell}^{(n)}(\omega_n) - i\hbar^{-1} \times \left[\mathcal{H}_{coh}(\omega_1), \rho^{(n-1)}(\omega_{n-1})\right]_{k\ell} \tag{2-25}$$

There is thus a systematic algebraic procedure to find all the Fourier components in the components ρ as a function of a power series in the perturbation. With the relation (2-7) for the expectation value of a physical quantity O, which may itself have an explicit time dependence, the Fourier components for the steady state response of each physical quantity may be found.

This procedure will now be illustrated with a very simple physical example. The Fourier components of the expectation value of the electric dipole moment at the second- and third-harmonic frequency will be calculated for an ensemble of localized one-electron systems. The coherent perturbation is assumed to be of the form

$$\mathcal{H}_{coh}(t) = -ex\{E_x \exp(-i\omega t) + E_x^* \exp(i\omega t)\} \tag{2-26}$$

The dipole moment operator ex is assumed to have only real off-diagonal elements $x_{ng} = x_{gn}$, $x_{nn} = 0$.

The only linear Fourier components are according to Eq. (2-23),

$$\rho_{ng}^{(1)}(\omega) = \frac{-\hbar^{-1} ex_{ng} E_x e^{-i\omega t}}{\omega - \omega_{ng} + i\Gamma_{ng}} \left(\rho_{gg}^{(0)} - \rho_{nn}^{(0)}\right) \tag{2-27}$$

and a similar expression with ω replaced by $-\omega$ and E_x by E_x^*. The expectation value of the linear electric dipole moment is

$$\begin{aligned}\langle \mathcal{P}_{(\omega)}^{(1)} \rangle &= \sum_g \sum_n \{ex_{gn}\rho_{ng}^{(1)}(\omega)\} \\ &= \sum_n \sum_g \frac{|ex_{gn}|^2 \hbar^{-1} 2\omega_{ng}}{\omega_{ng}^2 - (\omega + i\Gamma_{ng})^2} \rho_{gg}^{(0)}(E_x e^{-i\omega t})\end{aligned} \tag{2-28}$$

The double sum represents the complex polarizability of the system. If no distinction need be made between the macroscopic field and the applied field and the number of systems per unit volume in the state $|g\rangle$ is N_0, one finds for negligible damping the Kramers-Heisenberg dispersion formula,

$$\chi'(\omega) = 2N_0 \hbar^{-1} \sum_n |ex_{gn}|^2 \omega_{ng} (\omega_{ng}^2 - \omega^2)^{-1}$$

Note that the expression (2-28) is real in the limit $\omega \to 0$. There are no steady state losses for a dc field in a bound charge model. This feature has been obtained by considering the damping of the matrix elements $\rho_{ng}(\pm\omega)$ and $\rho_{gn}(\pm\omega)$ separately. In the real representation where no distinction between positive and negative frequencies is made, one has to correct the damping terms of the transverse components, e.g., in the Bloch equations, so that relaxation occurs toward thermal equilibrium at the instantaneous value of the Hamiltonian $\mathcal{H}_0 + \mathcal{H}_{coh}(t)$. This point has been treated in detail by Van Vleck and Weisskopf.[3] It is of somewhat academic value in the optical range, because the damping at an antiresonant term for which ω and ω_{ng} have the same sign will always be negligible. It is nevertheless gratifying that our approach gives the correct answer even at very low frequencies. Also note that the imaginary part correctly changes sign with ω, so that the sign of absorbed power is the same for positive and negative frequencies.

Consider next the lowest order nonlinear polarization. There will be Fourier components at the second-harmonic frequency $\pm 2\omega$ and at zero frequency. The change in population $\rho_{nn}^{(2)}(o)$ corresponds to a saturation effect. Since the diagonal elements of x_{nn} are assumed to be zero, only the off-diagonal elements $\rho_{nn}^{(2)}(2\omega)$ have to be determined for the calculation of the second-harmonic polarization. Equation (2-24) yields immediately on substitution of Eq. (2-27)

$$\begin{aligned}\rho_{n'n}^{(2)}(2\omega) = &\sum_{n''} \frac{e^2 x_{n'n''} x_{n''n} E^2 e^{-2i\omega t}}{\hbar^2 (2\omega - \omega_{n'n} + i\Gamma_{n'n})(\omega - \omega_{n''n} + i\Gamma_{n''n})} \\ &\times \left(\rho_{n''n''}^{(0)} - \rho_{nn}^{(0)}\right) \\ &+ \sum_{n''} \frac{e^2 x_{n''n} x_{n'n''} E^2 e^{-2i\omega t}}{\hbar^2 (2\omega - \omega_{n'n} + i\Gamma_{n'n})(\omega - \omega_{n'n''} + i\Gamma_{n'n''})} \\ &\times \left(\rho_{n''n''}^{(0)} - \rho_{n'n'}^{(0)}\right)\end{aligned} \qquad (2\text{-}29)$$

The second-harmonic dipole moment is

$$\langle \mathcal{P}_x(2\omega)\rangle = \chi_{xxx}(2\omega) E^2 e^{-2i\omega t} = \sum_n \sum_{n'} e x_{nn'} \rho_{n'n}(2\omega) \qquad (2\text{-}30)$$

In the case of negligible damping the nonlinear susceptibility may be written in another form by first combining the terms in $\rho^{(0)}_{n''n''}$ appearing in Eq. (2-29). Note that the term in the denominator $2\omega - \omega_{n'n}$ is just cancelled. This does not happen, when damping cannot be ignored because in general $\Gamma_{n'n} \neq \Gamma_{n''n} + \Gamma_{n'n''}$. In the absence of damping the susceptibility may be written after a relabeling of the indices in the triple summation as,

$$\chi'_{xxx}(2\omega) = \sum_{g} \sum_{\neq n} \sum_{\neq n'} N_0 e^3 x_{gn} x_{nn'} x_{n'g} \rho^{(0)}_{gg} \hbar^{-2}$$
$$\times \left[(2\omega - \omega_{ng})^{-1}(\omega - \omega_{n'g})^{-1} + (2\omega + \omega_{ng})^{-1} \right. \tag{2-31}$$
$$\left. \times (\omega + \omega_{n'g})^{-1} - (\omega + \omega_{ng})^{-1}(\omega - \omega_{n'g})^{-1}\right]$$

It is seen that this nonlinearity vanishes if the system has inversion symmetry and lacks accidental degeneracies. In this case the wave functions g, n, n′ have a definity parity and x can only connect states with opposite parity. Since x occurs on odd number of times the expression Eq. (2-31) will vanish. Due to our definition of the complex amplitudes this nonlinear susceptibility is a factor two larger than given in previous papers.[4] It is seen that the density matrix approach streamlines the elementary laborious calculations of the perturbed wave functions in section 2 in Appendix 1. At the same time the damping can be included in a natural way. Absorption and dispersion appear in the same order in the density matrix calculation. This approach is therefore used here and in Appendix 3.

There the reader will find complete expressions for the nonlinear polarization at the sum frequency $\omega_3 = \omega_1 + \omega_2$ when two periodic electric fields are applied. The symmetric form (see Eq. (2-23) of Appendix 3) only holds in the lossless case. The permutation symmetry relations derived from phenomenological energy considerations in Chapter I are thus confirmed by explicit calculation.

In a similar manner the third-harmonic polarization may be calculated from

$$\rho^{(3)}_{n''n}(3\omega) = \frac{-\sum_{n'} \left(e x_{n''n'} E^{(\omega)}_x \rho^{(2)}_{n'n}(2\omega) - \rho^{(2)}_{n''n'}(2\omega) e x_{n'n} E^{(\omega)}_x\right)}{\hbar(3\omega - \omega_{n''n'} + i\Gamma_{n''n'})} \tag{2-32}$$

In the summation n′ may be taken equal to n″ or n and we need also the Fourier components of the diagonal element $\rho^{(2)}_{nn}(2\omega)$, ignoring the small diagonal damping,

$$\rho_{gg}^{(2)}(2\omega) = \sum_n \frac{e^2 x_{ng}\, x_{gn}\, E^2 e^{-2i\omega t}}{\omega\hbar^2} \times \rho_{gg}^{(0)} \left[\frac{1}{\omega - \omega_{ng} + i\Gamma_{ng}} + \frac{1}{\omega + \omega_{ng} + i\Gamma_{ng}} \right] \tag{2-33}$$

Substitution of Eqs. (2-33) and (2-29) into (2-32) will give $\rho_{n''n}^{(3)}(3\omega)$ from which the third-harmonic polarization may be calculated

$$P(3\omega) = \chi(3\omega)\, E^3 e^{-3i\omega t} = N_0 \sum_{n''} \sum_{n} e x_{nn''}\, \rho_{n''n}^{(3)}(3\omega)$$

This should reduce to Eq. (2-22) in Appendix 1, if the damping is negligible and the change in populations can be ignored. This will always be so in practice. If the damping is negligible, several terms may be recombined and relabeled. One finds

$$\chi_{xxxx}^{(3\omega)} = \sum_{n''\neq g} \sum_{n'} \sum_{n\neq g} \sum_{g} \hbar^{-3} N_0 e^4 x_{gn}\, x_{nn'}\, x_{n'n''}\, x_{n''g} [\ \ ]\, \rho_{gg}^{0} \tag{2-34}$$

with

$$\begin{aligned}[\ \ ] = {} & (3\omega - \omega_{n''g})^{-1}(2\omega - \omega_{n'g})^{-1}(\omega - \omega_{ng})^{-1} \\ & - (3\omega + \omega_{n''g})^{-1}(2\omega + \omega_{n'g})^{-1}(\omega + \omega_{ng})^{-1} \\ & + (\omega - \omega_{n''g})^{-1}(2\omega - \omega_{n'g})^{-1}(\omega + \omega_{ng})^{-1} \\ & - (\omega + \omega_{n''g})^{-1}(2\omega + \omega_{n'g})^{-1}(\omega - \omega_{ng})^{-1}\end{aligned}$$

This result is identical with Eq. (2-22) of Appendix 1, except for a factor 4, which is due to our changed definition for the complex field amplitude.

The density matrix method has been used by several other authors to calculate the nonlinear response at optical frequencies.[5-10] The lowest order nonlinearity will be treated more systematically with two frequencies in the incident field and with retention of all multipole moments. This description is especially useful when the electron wave functions are not localized.

2-4 LOWEST ORDER NONLINEAR CONDUCTIVITY

Consider a solid in the one-electron approximation. Bloch wave functions will be used. Since the electrons probe effectively the spatially averaged electric and magnetic fields, no distinction need be made between the acting field and the macroscopic field. This distinction is important for localized electrons in dense media. Our discussions for localized electrons in optically dense materials must be modified. This will be done in the next chapter. More detailed considerations of Nozières and Pines[11] and Adler[12] confirm that the effective field is equal to the macroscopic field for Bloch electrons.

The electromagnetic field is derived from a vector potential

$$\mathbf{A}(\mathbf{r},t) = 2\mathrm{Re}\{\mathbf{A}_1 \exp(i\mathbf{q}_1 \cdot \mathrm{r} - i\omega_1 t) + \mathbf{A}_2 \exp(i\mathbf{q}_2 \cdot \mathbf{r} - i\omega_2 t)\} \tag{2-35}$$

In this section only the wave vector of the field will be denoted by $\mathbf{q}$, since $\mathbf{k}$ will be reserved for the wave vector of the Bloch electron. The Coulomb gauge will be used, in which the scalar potential vanishes, $\varphi = 0$.

$$\mathbf{E} = -\frac{1}{c}\frac{\partial \mathbf{A}}{\partial t} \qquad \mathbf{H} = \nabla \times \mathbf{A} \tag{2-36}$$

The interaction Hamiltonian for one nonrelativistic electron is

$$\mathcal{H}_{\mathrm{coh}} = -\frac{e}{2mc}(\mathbf{p}\cdot\mathbf{A} + \mathbf{A}\cdot\mathbf{p}) - \frac{e\hbar}{mc}\,\mathbf{s}\cdot\nabla\times\mathbf{A} + \frac{e^2}{2mc^2}\,\mathbf{A}\cdot\mathbf{A} \tag{2-37}$$

The current density operator for the electron with charge +e is

$$\begin{aligned} \mathbf{j}(\mathbf{r}_0) = +\frac{e}{2m}\Big\{&\Big[\delta(\mathbf{r}-\mathbf{r}_0)\Big(\mathbf{p}-\frac{e}{c}\mathbf{A}\Big) + \Big(\mathbf{p}-\frac{e}{c}\mathbf{A}\Big)\delta(\mathbf{r}-\mathbf{r}_0)\Big] \\ &+ 2i[\delta(\mathbf{r}-\mathbf{r}_0)(\mathbf{p}\times\mathbf{s}) - (\mathbf{p}\times\mathbf{s})\delta(\mathbf{r}-\mathbf{r}_0)]\Big\} \end{aligned} \tag{2-38}$$

where the position operator is defined by

$$\delta(\mathbf{r}-\mathbf{r}_0) = |\mathbf{r}_0\rangle\langle\mathbf{r}_0|$$

In the following the spin current density will be ignored and the spin term will also be omitted from the Hamiltonian. We wish to calculate the component of the current density at $\omega_3 = \omega_1 + \omega_2$. It is seen from Eq. (2-35) that this component will have a spatial variation exp $i(\mathbf{q}_1 + \mathbf{q}_2)\cdot\mathbf{r}$. Our problem is not only periodic in time but also periodic in space. Therefore a Fourier transform to wave vector space is made according to

$$\mathbf{j}(\mathbf{q},t) = \int \mathbf{j}(\mathbf{r},t)\, e^{-i\mathbf{q}\cdot\mathbf{r}_0}\, d\mathbf{r}_0 \tag{2-39}$$

Since $\mathbf{j}$ contains the time explicitly through $\mathbf{A}(t)$, three Fourier components of this operator in the frequency domain should be dintinguished.

$$\mathbf{j}(\mathbf{q}_1 + \mathbf{q}_2, 0) = +\frac{e}{2m}\{\mathbf{p}\, e^{-i(\mathbf{q}_1 + \mathbf{q}_2)\cdot\mathbf{r}} + e^{-i(\mathbf{q}_1 + \mathbf{q}_2)\cdot\mathbf{r}}\,\mathbf{p}\}$$

$$\mathbf{j}(\mathbf{q}_1 + \mathbf{q}_2,\omega_1) = -\frac{e^2}{mc} A_1 e^{-i\mathbf{q}_2\cdot\mathbf{r}}$$

$$\mathbf{j}(\mathbf{q}_1 + \mathbf{q}_2,\omega_2) = -\frac{e^2}{mc} A_2 e^{-i\mathbf{q}_1\cdot\mathbf{r}} \tag{2-40}$$

The Fourier component at $\omega_3 = \omega_1 + \omega_2$ of the current density, quadratic in the field amplitudes, is given by

$$\begin{aligned}\langle j^{NL}(\mathbf{q}_1 + \mathbf{q}_2,\omega_3)\rangle = \sum_n\sum_{n'} \Big\{&\rho^{(2)}_{nn'}(\omega_3)\,\mathbf{j}_{n'n}(\mathbf{q}_1 + \mathbf{q}_2,0)\\ &+ \rho^{(1)}_{nn'}(\omega_1)\,\mathbf{j}_{n'n}(\mathbf{q}_1 + \mathbf{q}_2,\omega_2)\\ &+ \rho^{(1)}_{nn'}(\omega_2)\,\mathbf{j}_{n'n}(\mathbf{q}_1 + \mathbf{q}_2,\omega_1)\Big\}\end{aligned} \tag{2-41}$$

The matrix elements of the perturbation $\mathcal{H}_{coh}$ between Bloch wave functions

$$u^*_{n'k'}\exp(-i\mathbf{k}'\cdot r) \qquad \text{and} \qquad u_{nk}\exp(+i\mathbf{k}\cdot\mathbf{r})$$

are given by

$$\begin{aligned}\mathcal{H}_{nn'}(\omega_{1,2}) &= -\frac{e}{mc}\,\delta(\mathbf{k} - \mathbf{k}' - \mathbf{q}_{1,2})(nk\,|\,\mathbf{p} + \hbar\mathbf{k}' + \frac{\hbar}{2}\mathbf{q}_{1,2}\,|\,n'k')\cdot\mathbf{A}_{1,2}\\ \mathcal{H}_{nn'}(\omega_3) &= +\frac{e^2}{2mc^2}\,\delta(\mathbf{k} - \mathbf{k}' - \mathbf{q} - \mathbf{q}_2)(nk\,|\,n'k')\mathbf{A}_1\cdot\mathbf{A}_2\end{aligned} \tag{2-42}$$

where

$$(n'\mathbf{k}'\,|O|\,nk) = \frac{1}{V}\int_V d\mathbf{r}u^*_{n'k'}\,O\,u_{nk}$$

with the integration over a unit cell of volume V. The Fourier components of the density matrix occurring in Eq. (2-41) can now be calculated by means of the recurrence relation Eq. (2-25). The density matrix in zero order is given by the Fermi-Dirac Distribution for thermodynamic equilibrium

$$\rho_{nn} = f(W_n) = \frac{1}{e^{(W_n - E_F)/kT} + 1} \tag{2-43}$$

Collection of all terms in Eq. (2-41) finally leads to the explicit expression, which is identical with an expression given by Cheng and Miller (their Eq. (13) in reference 9) and one given by Ducuing,[13]

$$\langle j_\alpha(q_1 + q_2, \omega_3)\rangle = \frac{e^3 E_{1\beta} E_{2\gamma}}{2m^2 \omega_1 \omega_2} \sum \text{Perm}(\beta, q_1, \omega_1;\ \gamma, q_2, \omega_2)\,[\quad]$$

$$
\begin{aligned}
[\quad] = {} & \sum_{n'nk} \frac{\left(n,\mathbf{k}\,|\,p_\beta + \hbar k_\beta - \tfrac{1}{2}\hbar q_{1\beta}\,|\,n', \mathbf{k} - \mathbf{q}_1\right)\left(n', \mathbf{k} - \mathbf{q}_1\,|\,n,k\right)\delta_{\gamma\alpha}\left\{f\left(E_{n,k}\right) - f\left(E_{n',\mathbf{k}-\mathbf{q}_1}\right)\right\}}{\hbar\left(\omega_1 + \omega_{n'\mathbf{k}-\mathbf{q}_1;\,n,\mathbf{k}} + i\Gamma_{n'\mathbf{k}-\mathbf{q},n\mathbf{k}}\right)} \\
& + \sum_{n'nk} \frac{1}{2}\,\frac{\left(n,\mathbf{k}\,|\,n'\mathbf{k} - \mathbf{q}_1 - \mathbf{q}_2\right)\left(n'\mathbf{k} - \mathbf{q}_1 - \mathbf{q}_2\,|\,p_\alpha + \hbar k_\alpha - \tfrac{1}{2}\hbar q_{1\alpha} - \tfrac{1}{2}\hbar q_{2\alpha}\,|\,nk\right)\delta_{\beta\gamma}\left\{f\left(E_{nk}\right) - f\left(E_{n',\mathbf{k}-\mathbf{q}_1-\mathbf{q}_2}\right)\right\}}{\hbar\left(\omega_1 + \omega_2 - \omega_{n,k;n'\mathbf{k}-\mathbf{q}_1-q_2} + i\Gamma_{n',\mathbf{k}-\mathbf{q}_1-\mathbf{q}_2;\,n\mathbf{k}}\right)} \\
& - \frac{1}{m} \sum_{n,n',n'',\mathbf{k}} \left(n\mathbf{k}\,|\,p_\beta + \hbar k_\beta - \tfrac{1}{2}\hbar q_{1\beta}\,|\,n'\mathbf{k} - \mathbf{q}_1\right)\left(n'\mathbf{k} - \mathbf{q}_1\,|\,p_\gamma + \hbar \mathbf{k}_\gamma - \hbar \mathbf{q}_1 - \tfrac{1}{2}\hbar \mathbf{q}_2\,|\,n'', \mathbf{k} - \mathbf{q}_1 - \mathbf{q}_2\right) \\
& \times \frac{\left(n'',\mathbf{k} - \mathbf{q}_1 - \mathbf{q}_2\,|\,p_\alpha + \hbar k_\alpha - \tfrac{1}{2}\hbar q_{1\alpha} - \tfrac{1}{2}\hbar q_\alpha\,|\,n,\mathbf{k}\right)\left\{f\left(E_{n'',\mathbf{k}-\mathbf{q}_1-\mathbf{q}_2}\right) - f\left(E_{n',\mathbf{k}-\mathbf{q}_1}\right)\right\}}{\left(\hbar\omega_2 - \hbar\omega_{n'\mathbf{k}-q_1;n'',\mathbf{k}-\mathbf{q}_1-\mathbf{q}_2} + i\Gamma_{n'n''}\right)\left(\hbar\omega_1 - \hbar\omega_{n,\mathbf{k};n'',\mathbf{k}-\mathbf{q}_1-\mathbf{q}_2} + i\Gamma_{nn''} + \hbar\omega_2\right)}
\end{aligned}
\tag{2-44}
$$

+ complex conjugate of last term with the signs of ω_1, ω_2, q_1, and q_2 reversed.

Apply this general expression first to the free electron gas. There is only one energy band. The energy transfer to an electron in an electron-photon collision is negligible in the optical region,

$$\hbar\omega_{nk,nk'} = \frac{\hbar^2}{2m}\left\{k^2 - (\mathbf{k} - \mathbf{q}_{1,2})^2\right\} \ll \hbar\omega_{1,2}$$

If the damping is also small, the denominators in Eq. (2-44) become independent of **k**. For transverse electromagnetic waves propagating in the α-direction, q has only α-components. Since u_{nk} is constant for free electrons, the matrix element of $p_\beta + \hbar k_\beta - \frac{1}{2} q_{1\beta}$ appearing in the first term of Eq. (2-44) is equal to k_β. When the summation over all k_β is performed, terms with k_β and $-k_\beta$ cancel. For the same reason the last term in Eq. (2-44) vanishes. The remaining term is

$$j_\alpha(\omega_1 + \omega_2) = \frac{e^3 E_{1\beta} E_{2\gamma}}{2m^2 \omega_1 \omega_2}$$

$$\times \sum_{\mathbf{k}} \frac{\left\{k_\alpha - \frac{1}{2}\left(q_{1\alpha} + q_{2\alpha}\right)\right\}\left\{f\left(E_{\mathbf{k}}\right) - f\left(E_{\mathbf{k}-q_{1\alpha}-q_{2\alpha}}\right)\right\}\delta_{\beta\gamma}}{\omega_1 + \omega_2}$$

$$= \frac{-Ne^3 \delta_{\beta\gamma} E_{1\beta} E_{2\gamma}}{2m^2 c\, \omega_1 \omega_2} \tag{2-45}$$

where N is the number of conduction electrons. For $\omega_1 = \omega_2$, this expression is identical to the classical expression,

$$j(2\omega) - Ne\dot{z}(2\omega)$$

where $\dot{z}(2\omega)$ is given by Eq. (1-6), with $\Gamma = \tau^{-1} = 0$.

Return to the electric dipole approximation by putting $\mathbf{q}_1 = \mathbf{q}_2 = 0$ in Eq. (2-44). It is clear that a vanishing result would be obtained for free electrons in this approximation. A matrix element of the type

$$\langle n'', \mathbf{k} - \mathbf{q}_1 - \mathbf{q}_2 \,|\, p_\alpha + \hbar(k_\alpha - \tfrac{1}{2}q_{1\alpha} - \tfrac{1}{2}q_{2\alpha}) \,|\, n, k\rangle$$

in Eq. (2-44) may be replaced in this approximation by the interband matrix element for the vertical transition,

$$(n'', k \,|\, p_\alpha \,|\, n, k) \equiv p^\alpha_{n''n} = \int u^*_{n''k}\, p_\alpha u_{nk}\, d\mathbf{r} \tag{2-46}$$

The nonlinear current density takes the form,

$$\langle j_\alpha(\omega_3)\rangle = \sigma_{\alpha\beta\gamma}(\omega_3 = \omega_1 + \omega_2)\, E_\beta(\omega_1)\, E_\gamma(\omega_2) \tag{2-47}$$

with a nonlinear conductivity tensor element

$$\sigma_{\alpha\beta\gamma}(\omega_3 = \omega_1 + \omega_2) = \frac{\hbar^{-2} e^3}{m^3 c^2 \omega_1 \omega_2}$$

$$\times \sum \text{Perm}\,(\beta, \omega_1; \gamma, \omega_2)$$

$$\times \sum_{n,n',n''} (2\pi)^{-3} \int f(E_{nk})\, d^3k\, [\quad]$$

$$[\quad] = \frac{p^{\alpha}_{nn'} p^{\beta}_{n'n''} p^{\gamma}_{n''n}}{\{\omega_1 + \omega_2 - \omega_{n'n}(k) + i\Gamma_{n'n}(k)\}\{\omega_2 - \omega_{n'n''}(k) + i\Gamma_{n''n}(k)\}}$$

$$+ \frac{p^{\beta}_{nn'} p^{\gamma}_{n'n''} p^{\alpha}_{n''n}}{\{\omega_1 + \omega_2 + \omega_{n''n}(k) + i\Gamma_{n''n}(k)\}\{\omega_1 + \omega_{n'n}(k) + i\Gamma_{n'n}(k)\}}$$

$$+ \frac{p^{\gamma}_{nn'} p^{\alpha}_{n'n''} p^{\beta}_{n''n}}{\{\omega_1 + \omega_2 - \omega_{n''n}(k) + i\Gamma_{n''n}(k)\}\{\omega_2 + \omega_{n'n}(k) + i\Gamma_{n'n}(k)\}}$$

$$+ \frac{p^{\gamma}_{nn'} p^{\alpha}_{n'n''} p^{\beta}_{n''n}}{\{\omega_1 + \omega_2 - \omega_{n''n'}(k) + i\Gamma_{n''n'}\}\{\omega_1 - \omega_{n''n}(k) + i\Gamma_{n''n}(k)\}} \tag{2-48}$$

The summation over $\mathbf{k}$ has been replaced by an integration. The $\mathbf{k}$ dependence of all matrix elements in the term [ ] is understood. The integration over the bands may be regarded in a certain sense as an inhomogeneous broadening of the resonance frequencies $\omega_{n'n''}$. The bands n'' are assumed to be empty.

If the medium has inversion symmetry, the functions u can be defined in such a manner that $u_{n,k}(r) = u_{n,-k}(-r)$ and thus $p^{\alpha}_{n'n}(\mathbf{k}) = -p^{\alpha}_{nn'}(-\mathbf{k})$. $\sigma_{\alpha\beta\gamma}$ therefore vanishes again in this dipole approximation. It may be shown in the usual way that in this same approximation, where the vector potential can be considered constant over the unit cell,

$$\begin{aligned} \mathbf{v}^{\alpha}_{nn'} &= \dot{x}^{\alpha}_{nn'} = i\omega_{nn'} x^{\alpha}_{nn'} \\ &= m^{-1} p^{\alpha}_{nn'} - (e/mc)A_{nn'} = m^{-1} p^{\alpha}_{nn'} \end{aligned} \tag{2-49}$$

The off-diagonal momentum matrix elements may therefore be replaced by the corresponding dipole matrix elements. The conductivity is related to the susceptibility as follows in this same approximation,

$$\mathbf{j}^{NL}(\omega_3) = \boldsymbol{\sigma}\mathbf{E}\mathbf{E} = \frac{\partial \mathbf{P}}{\partial t} = -i\omega_3 \chi \mathbf{E}\mathbf{E} \tag{2-50}$$

Equation (2-48) must therefore be entirely equivalent to Eqs. (2-21) and (2-22) in Appendix 3 if that expression is summed over all electrons of the occupied band.

This equivalence result can be proved quite generally, as first shown by Goeppert-Mayer.[14] Consider the Hamiltonian for a one-electron system with dimensions small compared to the wave length

$$\mathcal{H} = \frac{\mathbf{p}^2}{2m} + V(\mathbf{x}) - \frac{e}{mc}\mathbf{p}\cdot\mathbf{A}(\mathbf{r},t) + \frac{e^2}{2mc^2}\mathbf{A}^2(\mathbf{x},t) \tag{2-51}$$

The corresponding Lagrangian is

$$\mathcal{L} = \tfrac{1}{2}m\dot{\mathbf{r}}^2 - V(\mathbf{r}) + \frac{e}{c}\dot{\mathbf{r}}\cdot\mathbf{A}(\mathbf{r},t)$$

A total time derivative may be added to this Lagrangian without changing the equations of motion of the system. The new Lagrangian is therefore equivalent with the original one,

$$\mathcal{L}' = \tfrac{1}{2}m\dot{\mathbf{r}}^2 - V(\mathbf{r}) - \frac{e}{c}\mathbf{r}\cdot\frac{d\mathbf{A}(r,t)}{dt} \tag{2-52}$$

The equivalent Hamiltonian is

$$\mathcal{H}' = \frac{p^2}{2m} + V(\mathbf{r}) + \frac{e}{c}\mathbf{r}\cdot\frac{d\mathbf{A}(\mathbf{r},t)}{dt} \tag{2-53}$$

The total time derivative may be replaced by the partial time derivative if the spatial variation is negligible. In the electric dipole approximation the equivalent Hamiltonian is

$$\mathcal{H}' = \frac{p^2}{2m} + V(\mathbf{r}) - e\mathbf{r}\cdot\mathbf{E} = \mathcal{H}_0 + \mathcal{H}^1_{coh} \tag{2-54}$$

Fiutak[15] has generalized this proof and he has shown that in higher approximation the equivalent Hamiltonian indeed corresponds to the multipole expansion

$$\mathcal{H}_{coh} = -\mathbf{P}\cdot\mathbf{E} - \mathbf{M}\cdot\mathbf{H} - \mathbf{Q}:\nabla E + \cdots \tag{2-55}$$

where $\mathbf{Q}$ is the quadrupole moment. This expression does not account precisely for the build-up of diamagnetic energy in the system. This difficulty is of no importance for the nonlinear properties under discussion.

This expansion is physically very useful if the wave functions are localized. Valence electrons of molecules in liquids or gases, in molecular groups in solids, and in localized paramagnetic ions fall into this category. One may now expand the density matrix in a combined power series of the amplitudes **E** and **H** and of $\nabla\mathbf{E}$. Expectation values of the electric dipole moment, magnetic dipole moment, and electric quadrupole moment can be expressed as a sum of Fourier components, each of which is a combined power series of the electric and magnetic field amplitudes and their gradients. The procedure is straightforward, but cumbersome. Magnetic dipole or electric quadrupole terms are responsible for second harmonic generation in crystals with inversion symmetry. This has been observed in calcite. Recently a complete tabulation of all quadratic terms for the electric dipole, magnetic dipole, and electric quadrupole has been given by Adler.[10]

2-5 RAMAN-TYPE NONLINEARITIES

The current density proportional to third powers in the field amplitudes could in principle be determined in the same manner. In view of its complexity and because the cubic terms have an observable magnitude only for electric dipole matrix elements, these higher order terms are treated only in this approximation. The third-harmonic polarization has already been given in section 2-3. If the system is subjected to fields at two frequencies, numerous combination tones are possible. Among the many terms only a few will be singled out, because they have a small resonant or near-resonant denominator. Denote by ω_V a vibrational resonant frequency of the system. Let the two applied fields be at the laser frequency ω_L and a frequency ω_S near $\omega_L - \omega_V$. The subscript s is used to suggest the frequency with a Stokes shift. Only those terms will be retained which have a denominator $\Delta = \omega_L - \omega_S - \omega_V + i\Gamma_V$. All other frequencies ω_L, ω_S, $\omega_L + \omega_S$, etc., are supposed to be far removed from the resonant frequencies of the system. The Raman-type effects have been calculated in Appendix 3 for systems with two and three quantum levels.

Here a simplified model, due to Shen, will be chosen that represents the physical situation for a molecule exhibiting the Raman effect in its essential features. Consider the electronic ground state $|g\rangle$ and one excited electronic state $|n\rangle$. The nuclear vibrations are expanded in normal modes. The ground vibrational state $|0\rangle$ and the first excited state $|1\rangle$ of only one normal mode will be considered. The Hamiltonian of interaction consists of the coherent perturbation by the two light waves acting on the electron and an electron orbit-nuclear vibrational interaction,

$$\mathcal{H} = -exE_L(\omega_L) - exE_S(\omega_S) + \mathcal{H}_{o,v} \qquad (2\text{-}56)$$

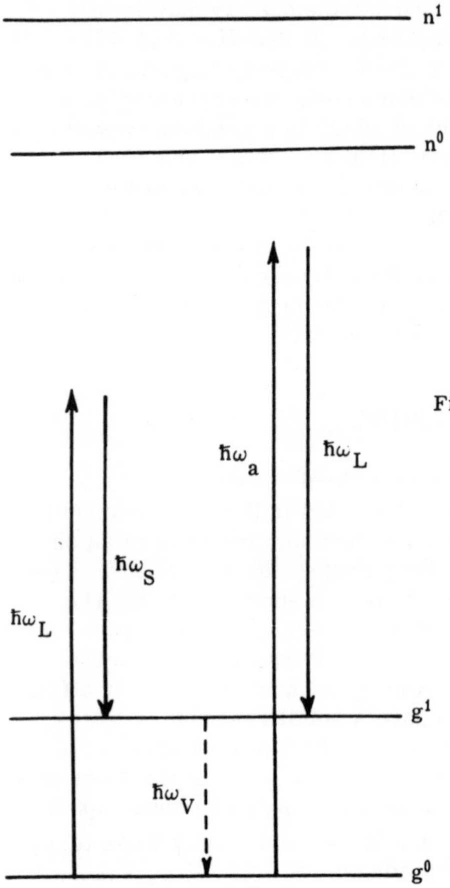

Figure 2-1. Raman-type transitions in a model with two electronic and two nuclear vibrational levels.

The four energy levels are shown in Figure 2-1. Due to the orbit-vibrational interaction the vibrational frequency is different in state $|g\rangle$ and $|n\rangle$. Assume that the only nonvanishing matrix elements are

$$(g^0|\mathcal{H}_{o,v}|g') \neq (n^0|\mathcal{H}_{o,v}|n')$$

and

$$(g^0|x|n^0) = (g'|x|n')$$

The interaction Hamiltonian has the matrix elements with Fourier components

$$\mathcal{H}_{n'g'}^{(\omega_s)} = \mathcal{H}_{ng}^{(\omega_s)} = -ex_{ng}E_s,$$

$$\mathcal{H}_{n'g^0}^{(\omega_s)} = \frac{(-ex_{ng})\left(\mathcal{H}_{ov}^{n'n^0} - \mathcal{H}_{ov}^{g'g^0}\right)}{\hbar\omega_v} E_s \approx \eta(-ex_{ng})E_s \tag{2-57}$$

$$\mathcal{H}_{n^0g'}^{(\omega_s)} = \frac{(-ex_{ng})\left(\mathcal{H}_{ov}^{n'n^0} - \mathcal{H}_{ov}^{g'g^0}\right)}{-\hbar\omega_v} E_s \approx -\eta(-ex_{ng})E_s$$

and similar expressions at ω_L. These equations are based on the Born-Oppenheimer approximation and the fact that the average vibrational separation $\hbar\omega_v = \frac{1}{2}\hbar(\omega_{g'g^0} + \omega_{n'n^0})$ is very small compared to the energy separation of electronic states, $\hbar\omega_{ng}$. The equations (2-57) express the fact that the dipole moment and consequently the electronic polarizability depends parametrically on the nuclear vibrational coordinate, as shown in detail by Placzek.[16]

The first order or linear approximation to the density matrix is, assuming that the molecule is initially in the ground vibrational state, $\rho_{g^0g^0} = 1$

$$\rho_{n^0g^0}^{(1)}(\omega_{L,s}) = \frac{\mathcal{H}_{n^0g^0}(\omega_{L,s})}{\hbar(\omega_{L,s} - \omega_{n^0g^0})} \qquad \rho_{n'g^0}^{(1)}(\omega_{L,s}) = \frac{\mathcal{H}_{n'g^0}(\omega_{L,s})}{\hbar(\omega_{(L,s)} - \omega_{n'g^0})} \tag{2-58}$$

In the second approximation one keeps only the terms with a resonant denominator

$$\rho_{g'g^0}^{(2)}(\omega_L - \omega_s) = \frac{\hbar^{-2}}{\omega_L - \omega_s - \omega_{g'g^0} + i\Gamma_{g'g^0}} \left\{ \frac{\mathcal{H}_{g'n^0}^{(-\omega_s)}\mathcal{H}_{n^0g^0}^{(\omega_L)}}{\omega_L - \omega_{n^0g^0}} + \frac{\mathcal{H}_{g'n'}^{(-\omega_s)}\mathcal{H}_{n'g^0}^{(\omega_L)}}{\omega_L - \omega_{n'g^0}} - \frac{\mathcal{H}_{g'n^0}^{(\omega_L)}\mathcal{H}_{n^0g^0}^{(-\omega_s)}}{+\omega_s + \omega_{n^0g^0}} - \frac{\mathcal{H}_{g'n'}^{(\omega_L)}\mathcal{H}_{n'g^0}^{(-\omega_s)}}{\omega_s + \omega_{n'g^0}} \right\} \tag{2-59}$$

The term $\rho^{(2)}_{n'n^0}(\omega_L - \omega_S)$ is either not at resonance, or it vanishes identically. From this expression the Fourier components

$\rho^{(3)}_{ng^0}(-\omega_S)$, $\rho^{(3)}_{n'g^0}(\omega_L)$, $\rho^{(3)}_{n'g^0}(2\omega_L - \omega_S)$ and $\rho^{(3)}_{n'g^0}(\omega_L - 2\omega_S)$

can in turn be determined. One finds, for example,

$$\rho^{(3)}_{n'g^0}(-\omega_S) = -\hbar^{-1}(+\omega_S + \omega_{n'g^0})^{-1}\mathcal{H}^{(-\omega_L)}_{n'g'}\rho^{(2)}_{g'g^0}(\omega_L - \omega_S)$$

$$\rho^{(3)}_{g'n'}(-\omega_S) = +\hbar^{-1}(\omega_L - \omega_{n'g^0})^{-1}\mathcal{H}^{(-\omega_L)}_{g^0n'}\rho^{(2)}_{g'g^0}(\omega_L - \omega_S)$$

$$\rho^{(3)}_{n^0g^0}(-\omega_S) = -\hbar^{-1}(+\omega_S + \omega_{n^0g^0})^{-1}\mathcal{H}^{(-\omega_L)}_{n^0g'}\rho^{(2)}_{g'g^0}(\omega_L - \omega_S)$$

$$\rho^{(3)}_{g'n^0}(-\omega_S) = +\hbar^{-1}(\omega_L - \omega_{n^0g^0})^{-1}\mathcal{H}^{(-\omega_L)}_{g^0n^0}\rho^{(2)}_{g'g^0}(\omega_L - \omega_S) \tag{2-60}$$

Finally the nonlinear polarization for N molecules per cubic centimeter

$$\langle P^{NL}(\omega_S)\rangle = \langle P^{NL}(-\omega_S)\rangle^* = N\,\mathrm{Tr}\,\{(ex)^*\rho^{*(3)}(-\omega_S)\}$$

is found to be

$$\langle P^{NL}(\omega_S)\rangle = \chi_S(\omega_S = \omega_L - \omega_L + \omega_S)\,|E_L|^2 E_S \tag{2-61}$$

with the "Stokes-susceptibility"

$$\chi_S = -\frac{Ne^4\hbar^{-3}\eta^2}{\omega_L - \omega_S - \omega_{g'g^0} - i\Gamma_{g'g^0}} \times \left|\frac{x_{ng}x_{gn}}{\omega_L - \omega_{n^0g^0}} - \frac{x_{ng}x_{gn}}{\omega_L - \omega_{n'g^0}} - \frac{x_{ng}x_{gn}}{\omega_S + \omega_{n^0g^0}} + \frac{x_{ng}x_{gn}}{\omega_S + \omega_{n'g^0}}\right|^2 \tag{2-62}$$

where η is defined by Eq. (2-57). Note that exactly at resonance, $\omega_L - \omega_S - \omega_{g'g} = 0$, χ_S is negative pure imaginary, corresponding to a negative absorption or positive gain at ω_S, proportional to the intensity of the laser beam $|E_L|^2$. In a similar manner one finds for the nonlinear polarization at ω_L,

$$P(\omega_L) = \chi_S^* E_L |E_S|^2 \tag{2-63}$$

There is a positive absorption at the frequency ω_L proportional to the intensity at ω_s. If the initial population of the vibrational state were inverted, $\rho^{(0)}_{g'g'} = 1$ instead of $\rho^0_{g^0 g^0} = 1$, the sign of χ_s would be inverted. The positive absorption would then be at ω_s and the gain at ω_L. Usually the excited vibrational state will not be populated appreciably.

The nonlinear polarization at the antistokes frequency, ω_a, caused by the field at ω_s and ω_L is given by

$$P(\omega_a) = \chi(\omega_a = 2\omega_L - \omega_s) E_L^2 E_s^* \tag{2-64}$$

Now the Hamiltonian (2-56) is augmented by another field at the antistokes frequency, E_a. One may calculate the nonlinear polarizations due to the Raman-type resonance process, at each of the three frequencies, ω_s, ω_L, and ω_a, in the simultaneous presence of three fields E_s, E_L, and E_a. Define an antistokes susceptibility in χ_a^*, obtained by replacing in Eq. (2-62), ω_L by ω_a and ω_s by ω_L. This quantity χ_a^* will only be slightly different from χ_s due to the denominators in the absolute square of Eq. (2-62). It can be shown by explicit calculation that the susceptibility in Eq. (2-64) may be expressed as $(\chi_s^* \chi_a)^{1/2}$. In the limit of negligible dispersion $(\chi_s^* \chi_a)^{1/2} = \chi_s^*$.

It should be remembered that there are other nonresonant contributions to the nonlinear susceptibilities. Their magnitude, although not enhanced by a resonant denominator, may nevertheless be appreciable because the matrix elements need not connect with nuclear vibrational levels. The reduction factor η does not occur in this part. The dispersion in this nonresonant contribution may be ignored because the spacing to excited electronic levels is much larger than the separation $\omega_a - \omega_s$. A single real nonresonant susceptibility χ_{NR} will suffice to describe this part of the nonlinear polarizations.

The results may be summarized in the following equations, which may be verified by explicit calculation of the same type that led to Eq. (2-62),

$$P(\omega_s) = (\chi_s + \chi_{NR}) |E_L|^2 E_s + \{(\chi_s \chi_a^*)^{1/2} + \chi_{NR}\} E_L^2 E_a^* \tag{2-65}$$

$$\begin{aligned} P(\omega_L) = {} & (\chi_s^* + \chi_{NR}) E_L |E_s|^2 + (\chi_a^* + \chi_{NR}) E_L |E_a|^2 \\ & + \{(\chi_s \chi_a^*)^{1/2} + (\chi_s^* \chi_a)^{1/2} + 2\chi_{NR}\} E_L^* E_s E_a \end{aligned} \tag{2-66}$$

$$P(\omega_a) = (\chi_a + \chi_{NR})|E_L|^2 E_a + \{(\chi_s^* \chi_a)^{1/2} + \chi_{NR}\} E_L^2 E_s^* \tag{2-67}$$

From Eqs. (2-67) we note the complex symmetry relationship,

$$(\chi_s \chi_a^*)^{1/2} = \chi(\omega_s = 2\omega_L - \omega_a) = \chi^*(\omega_a = 2\omega_L - \omega_s) \tag{2-68}$$

This is an illustration of the following rule which holds when the damping occurs in only one denominator. When the interchange in frequencies involves a change of sign of the frequency combination which happens to be near resonance, the complex conjugate should be taken. In $\chi(\omega_s = 2\omega_L - \omega_a)$ the near-resonance frequency appears as $\omega_L - \omega_a$, whereas in $\chi^*(\omega_a = 2\omega_L - \omega_s)$ the near-resonance frequency is $\omega_L - \omega_s = -(\omega_L - \omega_a)$. Therefore the complex conjugation is added. A detailed examination of the question how the damping term $i\Gamma_{n'n}$ is associated with the resonance $\omega_L - \omega_s - \omega_{n'n}$ will show the validity of this statement. One may surmise that this relation still holds for off-diagonal elements of the fourth-rank tensors if the indices are permuted with the frequencies,

$$\chi_{ijk\ell}(\omega_a = 2\omega_L - \omega_s) = \chi_{\ell jki}^*(\omega_s = 2\omega_L - \omega_a)$$

This is really an application of Onsager's relationships, which may be based on the existence of a dissipation function, describing the steady state loss. Pershan[17] has discussed this matter. He did not establish a relationship between the stored free energy and the dissipation function in the sense described here by the real and imaginary part of the same complex susceptibility.

In fact, Pershan and also Butcher and McLean[18] introduce complex susceptibilities or conductivities for lossless media. That part of the susceptibility which is described by tensors, in which an odd number of indices refers to a magnetic field or a magnetization, is purely imaginary. The part that is multiplied by an even number of magnetic field components is real in the lossless case. This follows from time-reversal symmetry arguments.

The pure electric dipole susceptibilities are therefore real in the nondissipative case. Their imaginary part describes losses. A symmetry relation, as in Eq. (2-68), represents therefore a connection between the free energy and the dissipation function. For the linear case this connection between the real and imaginary parts is embodied in the Kramers-Kronig relations. Our restriction that only those parts of the nonlinear susceptibility are considered with only one complex denominator implies that these parts obey the same

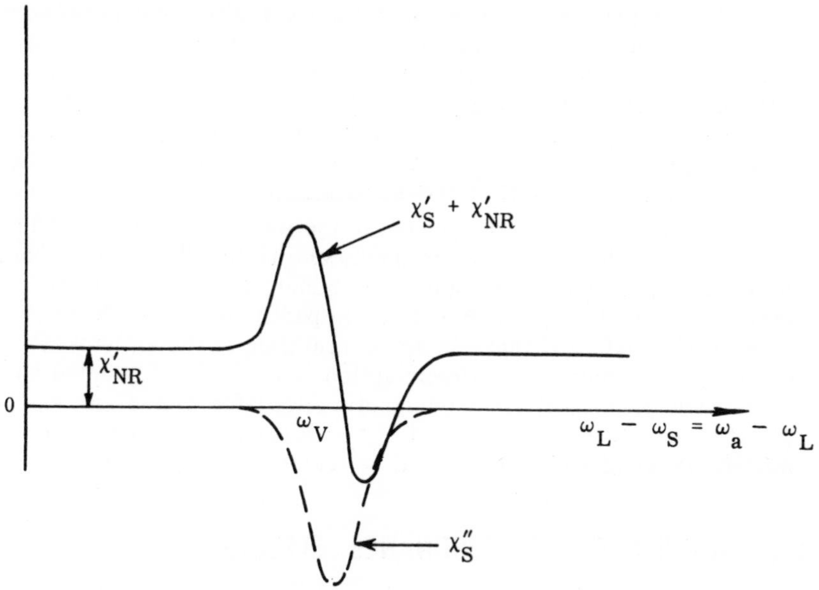

Figure 2-2. Dispersion and (negative) absorption of a complex Raman susceptibility. Note that the stokes frequency $\omega_s = \omega_L - \omega_v$ increases from right to left. The Kramers-Kronig relations are obeyed.

Kramers-Kronig relations as in the linear case, as shown in Figure 2-2. Only for these parts is the complex permutation symmetry relation Eq. (2-68) valid. The Raman-type susceptibilities fall into this category.

The imaginary part of the terms with χ_s and χ_a in Eqs. (2-67) describe the Raman processes shown in Figure 2-1. The real part of χ_s describes a parametric process, the simultaneous scattering of quanta at ω_L and ω_s, respectively. The interference of all these higher scattering processes in the homogeneous medium leads to a change in index of refraction at ω_s proportional to $|E_L|^2$, and vice versa.

The real part of the terms in $(\chi_s \chi_a^*)^{1/2}$ corresponds to a scattering process in which two quanta at ω_L scatter into a quantum at ω_s and one at ω_a, or vice versa. The interference of all scattering processes in the homogeneous medium leads to a parametric generation of ω_s and ω_L by the laser beam, or vice versa. The physical interpretation of the imaginary part of the terms in $(\chi_s \chi_a^*)^{1/2}$ and $(\chi_s^* \chi_a)^{1/2}$ is interesting. It can be described as the interference of two Raman processes, shown in Figure 2-1, operating between the same initial and the same final state. This interference term may either enhance or decrease the total transition rate between the vibrational levels. In the former case the generation of ω_s and the absorption at ω_a is enhanced; in the latter case the generation of ω_s and the absorption at ω_a is decreased by the interference. The relative phases of E_L, E_s, and E_a determine which situation applies. The very fact that the phases of the waves are important indicates that the interpretation in terms of quantum processes should not be taken seriously. The number of quanta is not uniquely determined if the phases are specified. In particular, it would be wrong to conclude that there is never real gain at the antistokes frequency. The only correct description is in terms of coupled coherent wave packets of electromagnetic oscillator states. An excellent approximation is afforded by the description in terms of coupled classical waves which will be treated in Chapter 4.

2-6 HIGHER ORDER RESONANCE EFFECTS

The expansion in terms of an ascending power series of the field amplitudes is appropriate under most circumstances in the optical region. Occasionally one would like to retain all powers of those field components which are near resonance. This situation is rather common in magnetic resonance at microwave and radio-frequencies. The most familiar example is the saturation effect between two levels at resonance,

$$\rho_{22} - \rho_{11} = \frac{\rho_{22}^0 - \rho_{11}^0}{1 + \hbar^{-2} | \mathcal{H}_{12}(\omega_{12})|^2 \Gamma_{12}^{-1} \Gamma_{11}^{-1}} \tag{2-69}$$

The zero frequency component of the diagonal matrix elements is expressed in terms of all powers of $(\mathcal{H}_{12}(\omega_{12})|^2$. One may always retain all powers of one Fourier component in the coherent perturbation by transformation to a rotating coordinate system, in which this component becomes time independent. This method has been worked out in detail in books on magnetic resonance.

If one can decide a priori by physical arguments of resonance which Fourier components of the density matrix elements will have a large response, all other Fourier components may be truncated off, i.e., put equal to zero, in the equation of motion (2-25) for the density matrix elements.

The steady state response is then described by a linear array of $\mathfrak{N}$ algebraic equations for the $\mathfrak{N}$ retained Fourier components. This procedure is illustrated for the two-level and three-level system in section III and IV of the Appendix 3. These particular components are thus obtained in all powers of the perturbation.

2-7 KRAMERS-KRONIG RELATIONS

The dispersion relations which exist for the linear susceptibilities have their analogies in the nonlinear case. Unfortunately, they are not very useful with the possible exception of second-harmonic generation. This was pointed out by Price[8] and this section will follow closely a discussion given by Caspers.[19]

The lowest order nonlinear steady state response of a system under the influence of a perturbation is given by the last term in Eq. (2-10) if the upper limits of integration are taken as ∞. For a perturbation $-\mathbf{P}\cdot(\mathbf{E}_1 e^{-i\omega_1 t} + \mathbf{E}_2 e^{-i\omega_2 t})$, one finds immediately for the steady state response at the sum frequency when a statistical average is taken over the ensemble and different damping histories,

$$\chi_{kij}(\omega_3^+ = \omega_1 + \omega_2) = -\frac{1}{2\hbar^2}\int_0^\infty dt \int_0^\infty d\tau\, e^{-i(\omega_1+\omega_2)t}$$

$$\Big\{e^{-i\omega_1\tau}\,\mathrm{Tr}\,[[\rho_0, \mathcal{P}_i(-t-\tau)], \mathcal{P}_j(-t)]\,\mathcal{P}_k + e^{-i\omega_2\tau}\,\mathrm{Tr}\,[[\rho_0, \mathcal{P}_j(-t-\tau)], \mathcal{P}_i(-t)]\,\mathcal{P}_k\Big\} \tag{2-70}$$

For $\omega_2 \geq \omega_1$ there will be a similar expression for $\omega_3^- = \omega_2 - \omega_1$ with ω_1 replaced by $-\omega_1$. Consider $\chi(\omega_3^{\pm}) = \chi(\omega_3)$ as a single function defined for both positive and negative values of ω_1. Because $\chi(\omega_3)$ appears in the general form $\int_0^\infty d\sigma\, e^{-\omega_1\sigma} F(\sigma,\omega_2)$, dispersion relations can formally be written down under very general conditions that χ is an analytic function in the lower half of the complex ω_1-plane,

$$\chi_{kij}(\omega_3) = \frac{i}{\pi}\,\mathrm{Pr}\int_{-\infty}^{+\infty} d\omega_1' \frac{\chi_{kij}(\omega_1',\omega_2)}{\omega_1' - \omega_1}, \quad \text{for} \quad \chi_{kij}(\infty,\omega_2) = 0 \tag{2-71}$$

where Pr stands for principal value.

The usefulness of a similar relation between the real and imaginary part of the linear susceptibility is that this relation may be split into two integrals, both going from 0 to $+\omega$, and that the real and imaginary parts of these integrals are separately of interest. This is not so in the nonlinear case. The function $\chi(\omega_3)$ must be separated into its physically significant parts $\chi(\omega_3^+ = \omega_2 + \omega_1' > 0)$, $\chi(\omega_3^- = \omega_1' - \omega_2 > 0$, and $\chi(-\omega_3^- = \omega_2 - \omega_1' > 0)$. One finds three integral relationships if the frequency axis ω_1' is cut into the appropriate parts,

$$\chi(\omega_3^+) = \frac{i}{\pi}\left[\Pr\int_0^\infty d\omega_1' \frac{\chi(\omega_3^+)}{\omega_1' - \omega_1} - \int_0^{\omega_2} d\omega_1' \frac{\chi(-\omega_3^-)}{\omega_1' + \omega_1}\right.$$
$$\left. - \int_{\omega_2}^\infty d\omega_1' \frac{\chi^*(\omega_3^-)}{\omega_1' + \omega_1}\right] \tag{2-72}$$
$$\chi(\omega_3^-) = \frac{i}{\pi}\left[\Pr\int_{\omega_2}^\infty d\omega_1' \frac{\chi(\omega_3^-)}{\omega_1' - \omega_1} + \int_0^{\omega_2} d\omega_1' \frac{\chi^*(-\omega_3^- = \omega_2 - \omega_1')}{\omega_1' - \omega_1}\right.$$
$$\left. - \int_0^\infty d\omega_1' \frac{\chi^*(\omega_3^+ = \omega_1' + \omega_2)}{\omega_1' + \omega_1}\right]$$

and a similar expression for $\chi(-\omega_3^-)$ by interchanging ω_1 and ω_2. These equations connect the real part of one susceptibility at the sum frequency jointly with imaginary parts at the sum and difference frequency. In the case of a fixed ratio between ω_1 and ω_2 and in particular for $\omega_1' = \omega_2$, the equations are simplified. The real part of the susceptibility for second-harmonic generation is related to its imaginary part by

$$\chi'(2\omega) = \frac{1}{\pi}\Pr\int_0^\infty d\omega' \frac{\chi''(2\omega')}{\omega' - \omega}$$
$$\chi''(2\omega) = -\frac{1}{\pi}\Pr\int_0^\infty d\omega' \frac{\chi'(2\omega')}{\omega' - \omega} \tag{2-73}$$

These relations are still not as useful as in the linear case, although they have the same formal identity. The imaginary part does not correspond to absorption, but merely refers to that part of the second harmonic which has a phase $2\varphi_1 + \pi/2$, where φ_1 is the phase of the fundamental signal. One usually knows the second-harmonic intensity only at a few selected frequencies, where sufficient fundamental intensity is available from a laser. If the second-harmonic production, proportional to $|\chi(2\omega)|^2$, were known over a wide frequency interval, the dispersion relations would give the relative phase with respect to the fundamental at each frequency. This information is not of much interest.

2-8 QUANTIZATION OF THE FIELDS

The susceptibilities have been defined by a semiclassical treatment in which the electromagnetic field is described as a classical complex parameter. The amplitude and phase of the wave are treated as c-numbers. This approach is very useful, because situations are described where the relative phases of different waves are important. The number of quanta is large enough that phases can be defined. The interest is not primarily in a transition from one state of the electromagnetic field with a prescribed number of quanta in each electromagnetic oscillator mode to another in which the number of quanta in one or more of these oscillators has changed by one or two or three. Such processes can be described in terms of transition probabilities per unit time or cross sections for scattering or absorption. In our case the interest is in a quasi-stationary state of the electromagnetic fields, which develops as a consequence of the interference of many scattering processes. The two situations are, of course, related because they are both described in terms of the same matrix elements of the interaction Hamiltonian between electromagnetic fields and matter. This problem is not peculiar to nonlinear optics. It is well known in linear optics where the linear susceptibility or index of refraction is related to the cross section for Rayleigh scattering and absorption. This question is discussed briefly by Kramers[20] and Heitler.[21] This section is devoted to the same problem for nonlinear case.

Nonlinear Absorption and Scattering Processes

The transition probability for a two-quantum absorption process was calculated by Mayer[14] and a scattering process in which three quanta are involved was treated by Blaton[22] and Guttinger.[23] The vacuum field is written as a linear combination of traveling waves,

$$\mathbf{A}(\mathbf{r},t) = \sum_{\lambda} [q_\lambda(t)\mathbf{A}_\lambda(\mathbf{r}) + q_\lambda^*(t)\mathbf{A}_\lambda^*(\mathbf{r})] \tag{2-74}$$

with

$$\mathbf{A}_\lambda(\mathbf{r}) = \hat{a}_\lambda \sqrt{4\pi c^2}\, e^{i\mathbf{k}_\lambda\cdot\mathbf{r}}\, V^{-1/2} \tag{2-75}$$

and a dynamical variable

$$q_\lambda(t) = q_\lambda e^{-i\omega_\lambda t} \tag{2-76}$$

The variables q_λ and q_λ^* are not Hermitian and not canonical conjugates, but a transformation to canonical real variables is possible,

$$\begin{aligned} Q_\lambda &= q_\lambda + q_\lambda^* \\ P_\lambda &= \dot{Q}_\lambda = -i\omega_\lambda (q_\lambda - q_\lambda^*) \end{aligned} \tag{2-77}$$

The Hamiltonian of the vacuum oscillators is

$$\mathcal{H}_\lambda = 2\omega_\lambda^2 q_\lambda q_\lambda^* = \tfrac{1}{2}(P_\lambda^2 + \omega_\lambda^2 Q_\lambda^2) \tag{2-78}$$

The mode wave functions obey the orthogonality relation

$$\int \mathbf{A}_\lambda \cdot \mathbf{A}_\mu^* \, dV = \int \mathbf{A}_\lambda \cdot \mathbf{A}_\mu \, dV = 4\pi c^2 \delta_{\lambda\mu} \tag{2-79}$$

The q_λ and q_λ^* are the raising (emission) and lowering (absorption) operators, respectively. They have the only nonvanishing matrix elements

$$\begin{aligned} q_{n_\lambda, n_\lambda+1} &= \left(\frac{\hbar(n_\lambda + 1)}{2\omega_\lambda}\right)^{1/2} e^{-i\omega_\lambda t} \\ q^*_{n_\lambda+1, n_\lambda} &= \left(\frac{\hbar(n_\lambda + 1)}{2\omega_\lambda}\right)^{1/2} e^{+i\omega_\lambda t} \end{aligned} \tag{2-80}$$

and obey the commutation relation

$$q_\lambda q_\mu^* - q_\mu^* q_\lambda = \frac{\hbar}{2\omega_\lambda} \delta_{\mu\lambda} \tag{2-81}$$

Let $|a\rangle$ and $|b\rangle$ represent two nondegenerate eigenstates of the material Hamiltonian. The interaction Hamiltonian (2-37) between the material system and the field then has the following nonvanishing matrix elements,

$$\begin{aligned} \mathcal{H}'(a, n_\lambda, b, n_\lambda + 1) &= -\frac{e}{mc}\left(\frac{2\pi\hbar c^2}{\omega_\lambda}\right)^{1/2} (n_\lambda + 1)^{1/2} e^{-i\omega_\lambda t} \\ &\quad \times \frac{1}{V^{1/2}} \int \psi_a^* (\mathbf{p} \cdot \hat{a}_\lambda) e^{i\mathbf{k}_\lambda \cdot \mathbf{r}} \psi_b \\ \mathcal{H}'(a, n_\lambda + 1, b, n_\lambda) &= -\frac{e}{mc}\left(\frac{2\pi\hbar c^2}{\omega_\lambda}\right)^{1/2} (n_\lambda + 1)^{1/2} e^{+i\omega_\lambda t} \\ &\quad \times \frac{1}{V^{1/2}} \int \psi_a^* (\mathbf{p} \cdot \hat{a}_\lambda) e^{-i\mathbf{k}_\lambda \cdot \mathbf{r}} \psi_b \end{aligned} \tag{2-82}$$

The term in A^2 of Eq. (2-37) with

$$A^2 = \sum_\lambda \sum_\mu (q_\lambda q_\mu \mathbf{A}_\lambda \cdot \mathbf{A}_\mu + q_\lambda q_\mu^* \mathbf{A}_\lambda \cdot \mathbf{A}_\mu^* + q_\lambda^* q_\mu \mathbf{A}_\lambda^* \cdot \mathbf{A}_\mu + q_\lambda^* q_\mu^* \mathbf{A}_\lambda^* \cdot \mathbf{A}_\mu^*) \tag{2-83}$$

contributes matrix elements

$$\mathcal{H}''(a, n_\lambda + 1, n_\mu; b, n_\lambda, n_\mu + 1) = \frac{e^2}{m} (\hat{a}_\lambda \cdot \hat{a}_\mu) \frac{2\pi\hbar}{(\omega_\lambda \omega_\mu)^{1/2}} (n_\lambda + 1)^{1/2}$$

$$\times (n_\mu + 1)^{1/2} e^{+i\omega_\lambda t - i\omega_\mu t}$$

$$\times \frac{1}{V} \int \psi_a^* e^{-i(\mathbf{k}_\lambda - \mathbf{k}_\mu)\cdot \mathbf{r}} \psi_b \tag{2-84}$$

and similar terms for

$$\mathcal{H}'(a, n_\lambda + 1, n_\mu + 1; b, n_\lambda n_\mu)$$

and

$$\mathcal{H}'(a, n_\lambda, n_\mu; b, n_\lambda + 1, n_\mu + 1)$$

The second-order matrix element representing the scattering of a quantum from the oscillator λ into the oscillator μ, while the material system goes from $|a\rangle$ to $|b\rangle$, is given by

$$\mathcal{H}_R = \sum_c \left\{ \frac{\mathcal{H}'(b, n_\lambda n_\mu + 1; c, n_\lambda n_\mu)\, \mathcal{H}'(c, n_\lambda n_\mu; a, n_\lambda + 1, n_\mu)}{-\hbar\omega_{ca} + \hbar\omega_\lambda} \right.$$

$$\left. + \frac{\mathcal{H}'(b, n_\lambda n_\mu + 1; c, n_\lambda + 1, n_\mu + 1)\mathcal{H}'(c, n_\lambda + 1, n_\mu + 1; a, n_\lambda + 1, n_\mu)}{-\hbar\omega_{ca} - \hbar\omega_\mu} \right\} + \mathcal{H}'' \tag{2-85}$$

If an integration is performed over a distribution of final states of the material, a time-proportional transition probability for the stimulated Raman scattering process is found by the "golden rule,"

$$w_R = \hbar^{-2} |\mathcal{H}_R|^2 g(\nu_\lambda - \nu_\mu) \tag{2-86}$$

The line shape function for the distribution of resonant frequencies due to damping by other random perturbations on the material system is a normalized Lorentzian

$$g(\nu) = \frac{2\Gamma_{ab}}{(\omega_{ba} - \omega)^2 + \Gamma_{ab}^2}, \quad \nu = \omega/2\pi \tag{2-87}$$

Assume that both waves are polarized in the x-direction to simplify the notation and that the resonance condition $\omega_{ba} = \omega_\lambda - \omega_\mu$

is satisfied. In the electric dipole approximation the probability per unit time for stimulated Raman scattering becomes at resonance

$$w_R = \frac{8\pi^2 e^4 \omega_\lambda \omega_\mu}{\Gamma_{ab} \hbar^2} (n_\lambda + 1)(n_\mu + 1) \times \frac{1}{V^2} \left| \sum_c \left(\frac{x_{bc} x_{ca}}{\omega_\lambda - \omega_{ca}} + \frac{x_{bc} x_{ca}}{-\omega_\mu - \omega_{ca}} \right) \right|^2 \tag{2-88}$$

A scattering cross section may be defined as follows. The number of initial quanta $n_\lambda + 1$ is contained in a volume V. The number of incident quanta per centimeter square per second is contained in a volume $V = c$. The transition probability for one quantum incident per centimeter square per second is the scattering cross section

$$\sigma_{coh}^{Raman}(\omega_\lambda) = \frac{8\pi^2 e^4 \omega_\lambda \omega_\mu \tilde{n}_\mu}{\hbar^2 \Gamma_{ab} c^2} \left| \sum_c \right|^2 \tag{2-89}$$

where $\tilde{n}_\mu$ is the number of incident quanta per centimeter square per second at the stokes frequency ω_μ.

A scattering cross section is directly related to the attenuation of the intensity per unit length, which in turn may be related to the imaginary susceptibility. Assume a dilute medium with N_o molecules per centimeter cubed and an index of refraction close to unity The argument could easily be extended to dense media. The linear absorption cross section σ_{abs} corresponds to an attenuation of the intensity according to

$$I(z) = I(o) \exp(-N_o \sigma_{abs} z) = I(o) \exp(-4\pi(\omega/c) \chi_1'' z)$$

In a similar manner the attenuation of the beam at ω_λ identified with the laser frequency, due to stimulated Raman scattering, is related to the out-of-phase part of the nonlinear Raman polarization

$$N_o \sigma_{coh}^{Raman} = 4\pi(\omega_\lambda/c)(-\chi_S'') \, |E_\mu|^2 \tag{2-90}$$

The number of incident quanta n_μ at the stokes frequency ω_μ leads to a power flux density

$$\frac{c}{2\pi} |E_\mu|^2 = n_\mu h\omega_\mu \tag{2-91}$$

The expression for the Poynting vector contains $\frac{1}{2}\pi$ rather than $\frac{1}{8}\pi$

because of our definition of the complex amplitudes. Combination of Eqs. (2-89), (2-90), and (2-91) finally gives the imaginary part of the Raman susceptibility at resonance.

$$(-\chi_S'') = \frac{N_0 e^4}{\Gamma_{ab}\hbar^3}\left|\sum_c\right|^2 \tag{2-92}$$

This checks with the expression (2-62) if the summation over c includes only the two levels N^0 and n' of Figure 2-1 and if the denominator of Eq. (2-62) is at resonance, $\omega_L - \omega_S = \omega_{g'g^0}$, which is identical to $\omega_\lambda - \omega_\mu = \omega_{ba}$. The levels c are assumed *not* to be at resonance for any of the frequencies involved. Note that this condition is necessary for the validity of Eq. (2-62) as well as of Eq. (2-85) and consequently of Eq. (2-92). There is therefore the same correspondence between the coherent Raman cross section and the imaginary part of the Raman susceptibility as there is between the coherent single photon absorption cross section and the imaginary part of the linear susceptibility.

In a similar manner the case of two quanta absorption processes may be treated. Instead of Eq. (2-88) one has a nearly identical expression with $-\omega_\mu$ replaced by $+\omega_\mu$. The resonance condition is now of course $\omega_\lambda + \omega_\mu = \omega_{ba}$. If n_μ in Eq. (2-89) still represents the number of incident quanta per centimeter square per second at ω_μ, the factor $n_\mu + 1$ should be replaced by n_μ in Eq. (2-88). The double photon absorption cross section is of course related to the imaginary susceptibility defined by

$$P^{NL}(\omega_\lambda) = i\chi''(\omega_\lambda = \omega_\lambda + \omega_\mu - \omega_\mu)E_\lambda|E_\mu|^2$$

$$P^{NL}(\omega_\mu) = i\chi''(\omega_\mu = \omega_\mu + \omega_\lambda - \omega_\lambda)E_\mu|E_\lambda|^2$$

Note that these two imaginary parts are equal, since there is simultaneous absorption at both frequencies. This is in agreement with our rule of section 2-5, since the near-resonant frequancy $\omega_\lambda + \omega_\mu$ has the same sign in both susceptibilities.

The reader will have noticed that the (+1) in $(n_\mu + 1)$ was dropped in going from Eq. (2-88) to Eq. (2-89). This represents, of course, the spontaneous Raman emission, when the incident number of stokes quanta $n_\mu = 0$. This emission will occur into all modes μ with ω_μ near $\omega_\lambda - \omega_{ba}$. In this case the density of final states is not determined by $g(\nu)$, but by the density of vacuum field oscillators. The replacement of $g(\nu_\lambda - \nu_\mu)$ in Eq. (2-86) by the density of modes for one polarization, $Vc^{-3}\nu_\mu^2\, d\Omega$, leads to a differential spontaneous Raman cross section per unit solid angle $d\sigma/d\Omega$, per unit volume of material for one polarization.†

$$d\sigma^{\text{Raman}}_{\text{ab spont.}} / d\Omega = \frac{N_0 e^4 \omega_\lambda \omega_\mu^3}{\hbar^2 c^4} \left| \sum_c \right|^2 \qquad (2\text{-}93)$$

Comparison of Eqs. (2-92) and (2-89) with Eq. (2-93) leads to a relation between the coherent gain and the incoherent cross section for spontaneous emission.

Scattering Cross Sections and Nonlinear Susceptibilities

The connection between the real part of these susceptibilities and the cross section is not so straightforward. The difficulty is highlighted when one considers a susceptibility such as $P(\omega_3) = \chi(\omega_3 = \omega_1 + \omega_2) E_1 E_2$, or any susceptibility of odd tensor rank. The transition probability or scattering cross section always contains absolute square values and this will always lead to even rank susceptibilities. The difficulty is, of course, connected with the fact that the complete specification of the number of quanta in the initial and final states of the fields implies complete uncertainty about the phases.

This problem already arises in the determination of the real part of the linear susceptibility with quantized fields. An attempt is usually made to connect the linear dispersion with the Rayleigh scattering cross section. This cross section is obtained by replacing ω_μ by $\omega_{\lambda'} = \omega_\lambda$ in the Raman cross section and $\mathbf{k}_\mu$

---

†The normalization question in these calculations presents the following difficulty. The scattering cross section may be normalized per molecule. The wave function of the molecule, appearing in Eq. (2-82), is normalized to unity. For N_0 molecules per unit volume, the incoherent scattering cross section per unit volume is N_0 times larger. The wave function of the incident photon flux is, however, normalized to a volume $V = c$, because the interest is in the number of photons passing through one square centimeter per second. This is the volume V appearing in Eq. (2-82) or Eq. (2-88).

The cross section in the electric dipole approximation is of course most easily determined directly from the equivalent Hamiltonian (2-54). One may also transform the expression for the cross section in terms of the momentum operator $\mathbf{p}$, usually found in the textbooks, to the form of Eq. (2-88) if proper care is taken to account for the commutation properties of $\mathbf{p}$ and $\mathbf{r}$. Details of this calculation are given by Kramers. It has been assumed that the scattering takes place in a plane perpendicular to the polarization of the electric fields. Eq. (2-88) can easily be generalized to the case where the electric field vectors before and after scattering are not parallel.

by $\mathbf{k}_{\lambda'} \neq \mathbf{k}_\lambda$. The final state of the material system is now the same as the initial state and will be denoted by g rather than a. The intermediate states will be denoted by n. Initially there is assumed to be no quantum in the field oscillator $\mathbf{k}_{\lambda'}$. The differential Rayleigh scattering cross section for one molecule in the ground state g into a unit solid angle for one direction of polarization is consequently obtained by comparison with Eq. (2-93),

$$d\sigma_{sp}^{Rayleigh}/d\Omega = \frac{e^4\omega_\lambda^4}{\hbar^2 c^4}\left|\sum_n\left\{\frac{-(\mathbf{r}\cdot\hat{\mathbf{a}}_\lambda)_{gn}(\mathbf{r}\cdot\hat{a}^*_{\lambda'})_{ng}}{+\omega_\lambda - \omega_{ng}} + \frac{(\mathbf{r}\cdot\hat{a}^*_{\lambda'})_{gn}(\mathbf{r}\cdot\hat{a}_\lambda)_{ng}}{\omega_\lambda + \omega_{ng}}\right\}\right|^2 \tag{2-94}$$

It is then usual to say that the scattered radiation corresponds to the radiation field of the dipole moment induced in the molecule by the incident field with wave vector $\mathbf{k}_\lambda$ and polarization $\hat{a}_\lambda$. This dipole moment is therefore proportional to the sum between the absolute value brackets. This is of course correct although it does not give information about the phase of this dipole moment. It is also not possible to make a direct connection with the linear dispersion for another reason.

Consider first an assembly of N_0 molecules per unit volume, with position vector $\mathbf{R}_i$. Although the phases of the modes with a prescribed number of quanta are not known, the relative phase at different positions is determined by the spatial mode functions in Eq. (2-75). The Rayleigh cross section for this unit volume element is therefore,

$$\sigma^{Rayl} = \frac{e^4\omega_\lambda^4}{\hbar^2 c^4}\left|\sum_{i=1}^{N_0} e^{+i(\mathbf{k}_\lambda - \mathbf{k}_{\lambda'})\cdot\mathbf{R}_i}\sum_n\{\ \}\right|^2 \tag{2-95}$$

where the expression between curly brackets is the same as in Eq. (2-94). When the N_0 molecules are evenly spaced as in a crystal, the sum over i will vanish in the limit of a large volume, except for those directions $\mathbf{k}'_\lambda$, where the Bragg reflection condition for X-rays is satisfied. At optical frequencies this condition is never satisfied in dense media, since the optical wave length is much larger than the spacing between molecules. In homogeneous crystals and liquids there is no Rayleigh scattering, except that due to fluctuations in the density from the mean periodic or homogeneous distribution. There can of course be constructive interference of the radiation from the various induced dipoles in the forward direction with $\mathbf{k}_\lambda = \mathbf{k}_{\lambda'}$. It is, however, precisely in this direction that the scattering formula for light quanta (2-95) cannot be applied. The final state of the system consisting of atoms and field is then identical with the initial state. There is no transition probability. It is precisely the interference of the forward scattered wavelets with the incident wave that determines the index of refraction of the medium, and thus the path of light rays and many optical laws.

The same difficulty arises with the nonlinear susceptibilities. It would be possible to identify the nonlinear dipole moment of a material system lacking inversion symmetry, $\mathcal{P}(\omega_3) = \chi(\omega_3 = \omega_1 + \omega_2)E_1(\omega_1)E(\omega_2)$, with a scattering cross section for a process involving three quanta and three modes of the vacuum field $\mathbf{k}_1$, $\mathbf{k}_2$, and $\mathbf{k}_3$. The initial number of quanta in these three modes is $n_{k_1} + 1$, $n_{k_2} + 1$, 0. The transition probability to a final state with n_{k_1}, n_{k_2}, $n_{k_3} = 1$ is calculated. The radiation at ω_3 is identified with the mean square of a dipole moment at ω_3. By considering a volume element of a crystal containing a large number of unit cells, it may even be shown, in analogy to Eq. (2-95), that the total scattering matrix element contains a sum $\Sigma_i \exp i(\mathbf{k}_1 + \mathbf{k}_2 - \mathbf{k}_3)\cdot\mathbf{R}_i$. For a sufficiently large volume this sum vanishes, unless $\mathbf{k}_3 = \mathbf{k}_1 + \mathbf{k}_2$. This is the condition of conservation of momentum in the collision between three quanta. The lattice can only take up momenta proportional to the reciprocal lattice vector. This has the order of magnitude of X-ray momenta. Since the lattice cannot balance the momentum of visible light quanta, as a single atom can, the momentum must be conserved among the optical quanta in a large enough crystal.

When initially there is also a field E_3 present at ω_3, the semiclassical treatment has shown that it is of vital importance to know what the relative phase is of $\mathcal{P}(\omega_3)$ determined by the phases of E_1 and E_2 with respect to the phase of E_3. Whether the wave at E_3 will grow, i.e., do negative work on the medium, or decay by doing positive work on the medium, depends on this relative phase. If the phases are undetermined, as in the quantum treatment with a prescribed number of quanta, these features, sensitive to the relative initial phases of the incident waves, are lost.

Although the consideration of scattering processes, where the cross section has powers of the field amplitudes twice as high as the rank of the susceptibility tensor sought, can give much information, some phase senstive features are irretrievably lost. The subtle interplay between real and imaginary parts of the complex linear and nonlinear susceptibilities follows quite naturally from the semiclassical treatment. How can this information be obtained from a theory, in which the fields are quantized?

Coherent Quantum States. Limitation of the Semiclassical Treatment

It will obviously be necessary to put the phase information of the initial states of the vacuum field into the quantum description, which should follow the classical description of the waves as closely as possible. This may be done by introducing minimum uncertainty wave packets in the Ehrenfest sense, consisting of a suitable superposition of quantum states with a prescribed number of quanta. This allows the best possible simultaneous definition of amplitude and phase in quantum mechanics. It can be shown that the Poisson distribution over states with different n is such a state for which

$$\langle \varphi^2 - \langle \varphi \rangle^2 \rangle^{1/2} \langle n^2 - \langle n \rangle^2 \rangle^{1/2} = \tfrac{1}{2}$$

$$\psi_{coh} = \sum_{n=0}^{\infty} c_n \, e^{-i\omega t(n+\frac{1}{2})} \, | n \rangle$$

$$c_n = \left(\frac{e^{-\bar{n}} (\bar{n})^n \, e^{-2in\varphi}}{n!} \right)^{1/2}, \quad \bar{n} = \sum_{n=0}^{\infty} |c_n|^2 \, n \tag{2-96}$$

These coherent states, already introduced by Schrödinger, have recently been investigated in much detail by Glauber[24] and many other workers. It would carry us too far to discuss the susceptibilities in terms of these states. The correspondence principle guarantees that in the limit of large $\bar{n}$ the results are identical with the semiclassical method. Nearly all nonlinear optical phenomena are observed in the domain of large $\bar{n}$. A typical Q-switched laser pulse contains about 10^{18} light quanta distributed perhaps over 10^2 to 10^4 modes. When a laser beam traverses a medium which has some loss at the laser frequency, the wave is exponentially attenuated. Classically the phase remains determined. The minimum uncertainty wave packet described by Eq. (2-96) will make transitions to other coherent states with a smaller value of $\bar{n}$. The phase information is retained as far as possible. It is of course true that the phase expectation value is not well determined when the beam has been attenuated so much that $\bar{n} < 10$ or 5.

Similarly, a build-up of intensity at the second-harmonic frequency has to start from the zero-point vibrations. The phase is initially not well determined, but any wave packet at frequency 2ω

with an appreciable value of $\langle n(2\omega)\rangle$ will have a phase $\langle \varphi(2\omega)\rangle$ close to $2\langle \varphi(\omega)\rangle - \pi/2$ according to the correspondence principle. For this phase the negative work done by the field on the nonlinear polarization with phase $2\langle \varphi(\omega)\rangle$ is extreme and there is maximum growth. Due to dispersion the phases of the wave packets will get out of step. All these results which follow easily from the semiclassical treatment should be reproduced by the coherent quantum states.

A rigorous description of the properties of quantum noise in parametric devices and lasers should be given in terms of these coherent states. Much work has been done on this question by Louisell,[25] Gordon[26,27] and coworkers.

The semiclassical theory which is used in this monograph will describe all situations correctly in a much simpler fashion with the exception of those low levels of intensity where quantum noise is important. In that case the semiclassical theory is augmented by an *ad hoc* introduction of the spontaneous emission process described earlier in this section. Fields, with amplitudes to give the appropriate intensity for spontaneous emission, and with random phases, are added in an *ad hoc* fashion to the coherent fields. These are the "noise sources" that are added to the classical fields. The classical field in a laser builds up from this noise. The phase of the field in a laser mode is not known a priori. As the wave builds up and is fed back by reflection, a definite, but unknown, phase is established. In a similar way the phase of the field at the stokes frequency in a Raman laser is not known a priori. It is believed that the combination of the semiclassical theory and the incoherent noise sources will give a satisfactory, although not rigorous, description of all situations of practical interest.

Quantum Theory of Damping

This chapter is concluded with an outline how the damping terms in the density matrix can be obtained for interactions with quantized random fields. The optical homogeneous line width is often due to spontaneous emission of photons or phonons. The phonon field may be quantized in the same manner as the electromagnetic field. Consider only two energy levels $|a\rangle$ and $|b\rangle$ of the material system Hamiltonian $\mathcal{H}_0$ to simplify the algebra. The Hamiltonian of the field (electromagnetic or vibrational) is denoted by $\mathcal{H}_f$. The interaction between the material system and the field is assumed to be given in the form of the product of an operator O acting on the material system and an operator F acting on the field variables. The stochastic time-dependent perturbation is

$$\mathcal{H}_{int}(t) = O\,F(t) = e^{+i\hbar^{-1}\mathcal{H}_f t}\, F\, e^{-i\hbar\mathcal{H}_f t}\, O \qquad (2\text{-}97)$$

The random function $F(t)$ is now a quantum operator acting on the field variables. The probability to find the system in the state $|a\rangle$ and the field in state $|f'\rangle$ at time t, when the system was in the state $|b\rangle$ and the field in state $|f\rangle$ at $t = 0$ is given according to Eq. (2-17) by,

$$W_{af',bf}(t) = \hbar^{-2}\int_0^t dt' \int_0^t dt'' \langle b,f|\mathcal{H}_{int}(t')|a,f'\rangle \times \langle a,f'|\mathcal{H}_{int}(t'')|bf\rangle$$

$$= |O_{ab}|^2\hbar^{-2}\int_0^t dt \int_{-\infty}^{\infty} d\tau \langle f|F(t)|f'\rangle \times \langle f'|F(t-\tau)|f\rangle e^{-i\omega_{ba}\tau}$$

This leads to a time-proportional transition probability when a statistical average is taken over the initial state f of the field and a summation is performed over the final states f′,

$$w_{b\to a} = \hbar^{-2}|O_{ab}|^2 \int_{-\infty}^{+\infty} d\tau \; e^{-i\omega_{ab}\tau} \times \sum_{f,f'} p(f)\langle f|F(t)|f'\rangle\langle f'|F(t-\tau)|f\rangle \tag{2-98}$$

The probability to find the system initially in the state f is

$$p(f) = \frac{e^{-E_f/kT}}{\Sigma\, e^{-E_f/kT}} = \frac{e^{-\mathcal{H}_f/kT}}{\mathrm{Tr}\, e^{-\mathcal{H}_f/kt}} \tag{2-99}$$

One may define the quantum mechanical analogue of a correlation function,

$$g(\tau) = \mathrm{Tr}\,[F(t)\,p(\mathcal{H}_f)\,F(t+\tau)] \tag{2-100}$$

independent of the representation of $\mathcal{H}_f$. With the corresponding spectral density,

$$J(\omega) = \int_{-\infty}^{+\infty} g(\tau)e^{-i\omega\tau}\, d\tau \tag{2-101}$$

the transition probability per unit time can be written

$$w_{b\to a} = \hbar^{-2}|O_{ab}|^2\, J(\omega_{ba}) \tag{2-102}$$

For the reverse transition one has

$$w_{a\to b} = \hbar^{-2}|O_{ab}|^2 J(\omega_{ab}) \tag{2-103}$$

These two expressions are not equal, as may most easily be seen by evaluating $J(\omega)$ in the representation in which $\mathcal{H}_f$ is diagonal.

$$\left(\sum e^{-E_F/hT}\right) J(\omega_{ab}) = \int_{-\infty}^{+\infty} \sum_{ff'} |\langle f|F|f'\rangle|^2 e^{-E_f/kT} e^{i(\omega_{ff'}-\omega_{ab})\tau} d\tau \tag{2-104}$$

Since the states in the reservoir are dense and uniformly distributed, the integral over τ is essentially a δ-function.

$$\left(\sum e^{-E_F/kT}\right)\cdot J(\omega_{ab}) = \sum_f |\langle f|F|f - \omega_{ab}\rangle|^2 e^{-E_f/kt}$$

$$\left(\sum e^{-E_F/kT}\right) J(-\omega_{ab}) = \sum_f |\langle f|F|f + \omega_{ab}\rangle|^2 e^{-E_f/kT}$$

$$= \sum_f |\langle f - \omega_{ab}|F|f\rangle|^2 e^{-E_f/kT} e^{+\omega_{ab}/kT}$$

Therefore

$$w_{a\to b} = w_{b\to a} e^{+\hbar\omega_{ab}/kT} \tag{2-105}$$

The a priori probability for the material system to make a transition downward in energy is higher than the transition upward, precisely in the way postulated in Eq. (2-20) to satisfy the principle of detailed balancing. The reason is of course simple. A downward transition in the material system requires a corresponding upward transition in the fields that serve as a thermal reservoir. Since the states with lower energy in the reservoir have a higher a priori probability to be occupied, upward transitions between a given pair of levels in the reservoir are more likely.

At optical frequencies $h\omega >> kT$, the reservoir temperature may be put equal to $T = 0$. The vacuum oscillators at light frequencies are in the ground state. There are no upward transitions in the material system. The downward transition corresponds to spontaneous emission. Eq. (2-104) may be written as

$$J(+\omega_{ab}) = \int_{-\infty}^{+\infty} \sum_{\lambda} |\langle n_\lambda = 0|F|n_\lambda = 1\rangle|^2 e^{i(\omega_\lambda - \omega_{ab})\tau} d\tau \tag{2-106}$$

where λ is a sum over all vacuum oscillators. The integration over

τ gives essentially the density of oscillators at the resonant frequency $c^{-3} V \nu_{ab}^2 \, d\nu \, d\Omega$. The spontaneous emission probability follows in the usual way from Eqs. (2-103) and (2-106) if the proper expression for the matrix element O_{ab} is substituted. All considerations of section 2-2 now apply to damping by spontaneous emission. If the effect of the finite temperature of the thermal radiation field is taken into account, one verifies immediately from Eqs. (2-104) and (2-106) that the transition probability downward is

$$w_{a \to b} = (\bar{n} + 1) \, w_{a \to b}^{sp} \tag{2-107}$$

and the transition probability upward is

$$w_{b \to a} = \bar{n} \, w_{a \to b}^{sp} \tag{2-108}$$

where $\bar{n} = [\exp(\hbar\omega_{ab}/kT) - 1]^{-1}$ is Planck's function for the average number of thermally excited photons.

The transverse relaxation of the two-level system is described by the constant Γ_{ab}. If there are no diagonal elements of the interaction Hamiltonian causing adiabatic contributions, the transverse damping is given by

$$\Gamma_{ab} = \tfrac{1}{2}(\Gamma_{aa} + \Gamma_{bb}) = \tfrac{1}{2}(2\bar{n} + 1) \, w_{a \to b}^{sp} \tag{2-109}$$

The frequency response to a harmonic signal is according to the considerations of section 2-3 a Lorentzian, which at low temperatures ($\bar{n} \ll 1$) takes the form

$$g(\nu) = \frac{w_{a \to b}^{sp}}{(\omega - \omega_{ab})^2 + \frac{1}{4}\left(w_{a \to b}^{sp}\right)^2}, \quad \nu = \omega/2\pi \tag{2-110}$$

This line shape function also describes the spontaneous emission line itself. It has this shape for emission in each direction for each polarization. The argument is as follows. Single out one vacuum oscillator in the specified direction, with the specified polarization, at frequency ω. All other vacuum oscillators still produce essentially the same random perturbation operator as before, consequently the same damping and the same line shape function $g(\nu)$ prevail. The treatment of spontaneous emission in this manner is clearly based on the large density of vacuum oscillators. The radiation field can indeed act as a thermal reservoir. The situation is, of course, radically different if a material system is enclosed in a cavity which allows only one, or a few modes, in the frequency range at interest.

Jaynes[28] has given an elegant quantum mechanical discussion of the interaction between a material system with two levels and one radiation oscillator. At optical frequencies this situation is not encountered.

This chapter on quantum mechanics is now concluded. The expectation values of the Fourier components of the polarization may be calculated for ascending powers in the applied field amplitudes by the density matrix method in the semiclassical approximation. The phenomenological damping terms are determined by random perturbations, which include the damping by spontaneous emission. The expectation values of the nonlinear polarization, determined by the incident fields, must now in turn be regarded as additional sources for these fields. This will be done in the next chapter. Since the Fourier components are calculated by a semiclassical method, the fields due to spontaneous emission must be added in an *ad hoc* manner, with random phases, to the classical fields.

REFERENCES

1. Excellent descriptions of the density matrix method and its applications may be found for example in: R. C. Tolman, *Principles of Statistical Mechanics*, Oxford University Press, New York, 1938; A. Abragam, *The Principles of Nuclear Magnetism*, Oxford University Press, New York, 1961. C. P. Slichter, *Principles of Magnetic Resonance*, Harper & Row, New York, 1962.
2. N. Bloembergen and Y. R. Shen, *Phys. Rev.*, **133**, A37 (1964). This paper is reprinted as Appendix 3. It contains many references to the literature. Compare also, V. M. Fain and E. Q. Yaschin, *J. Exp. Theor. Phys. (U.S.S.R.)*, **46**, 695 (1964), English translation, *JETP*, **19**, 474 (1964).
3. J. H. Van Vleck and V. F. Weisskopf, *Rev. Mod. Phys.* **17**, 227 (1945).
4. J. A. Armstrong, N. Bloembergen, J. Ducuing, and P. S. Pershan, *Phys. Rev.*, **127**, 1918 (1962). This paper is reprinted as Appendix 1.
5. P. N. Butcher and T. P. McLean, *Proc. Phys. Soc.* (London), **81**, 219 (1963).
6. R. Loudon, *Proc. Phys.* Soc. (London), **80**, 952 (1962).
7. P. L. Kelley, *J. Chem. Phys. Solids*, **24**, 607 (1963).
8. P. J. Price, *Phys. Rev.*, **130**, 1792 (1963).
9. H. Cheng and P. B. Miller, *Phys. Rev.*, **134**, A 683 (1964).
10. E. Adler, *Phys. Rev.*, **134**, A 728 (1964).
11. P. Nozieres and D. Pines, *Phys. Rev.*, **109**, 762 (1958).
12. S. Adler, *Phys. Rev.*, **126**, 413 (1962).
13. J. Ducuing, Thèse, Université de Paris (1964).
14. M. Goeppert-Mayer, *Ann. Physik*, **9**, 273 (1931).

15. J. Fiutak, *Can. J. Phys.*, **41**, 12 (1963). See also, E. A. Power and S. Zienau, *Phil. Trans. Roy. Soc.*, **251** A, 54 (1959).
16. G. Placzek, *Marx Handbuch der Radiologie*, 2nd ed., Vol. VI, Part II, (1934), pp. 209–374.
17. P. S. Pershan, *Phys. Rev.*, **130**, 919 (1963).
18. P. N. Butcher and T. P. McLean, *Proc. Phys. Soc.* (London), **83**, 579 (1964).
19. W. J. Caspers, *Phys. Rev.*, **133** A, 1249 (1964).
20. H. A. Kramers, *Quantum Mechanics*, trans. D. ter Haar, North-Holland Publishing, Amsterdam, 1957, p. 482 ff.
21. W. Heitler, *Quantum Theory of Radiation*, 3rd ed., Oxford University Press, New York (1954).
22. J. Blaton, *Z. Physik*, **69**, 835 (1931).
23. P. Guttinger, *Helv. Phys. Acta*, **5**, 237 (1932).
24. R. J. Glauber, *Phys. Rev.*, **130**, 2529 (1963); **131**, 2766 (1963).
25. W. H. Louisell, A. Yariv, and A. E. Siegman, *Phys. Rev.*, **124**, 1646 (1961).
26. J. P. Gordon, W. H. Louisell, and L. R. Walker, *Phys. Rev.*, **129**, 481 (1963).
27. J. P. Gordon, L. R. Walker, and W. H. Louisell, *Phys. Rev.*, **130**, 806 (1963).
28. E. T. Jaynes and F. W. Cummings, *Proc. IEEE*, **51**, 89 (1963).

3

MAXWELL'S EQUATIONS IN NONLINEAR MEDIA

The nonlinear polarizations or current densities have been expressed in terms of the fields. These quantities serve in turn as sources for the fields. This second step is discussed in this chapter. The nonlinear current density, calculated either classically by the methods of Chapter 1 or as the expectation value of a quantum mechanical operator by the methods of Chapter 2, must now be incorporated in Maxwell's equations. From a microscopic point of view one has only the vacuum equations and a collection of moving point charges

$$\nabla \times \mathbf{b} = \frac{1}{c}\frac{\partial \mathbf{e}}{\partial t} + \frac{4\pi}{c}\mathbf{j} \qquad \nabla \cdot \mathbf{e} = 4\pi\rho$$
$$\nabla \times \mathbf{e} = -\frac{1}{c}\frac{\partial \mathbf{b}}{\partial t} \qquad \nabla \cdot \mathbf{b} = 0 \tag{3-1}$$

Here $\mathbf{j}$ is the microscopic current density, both linear and nonlinear, including the spin current density, corresponding to Eq. (2-38). The macroscopic equations of Maxwell are obtained by an averaging process over a volume, large compared to the atomic dimensions, but small compared to the wave length.[1] This procedure is, of course, not useful at very short wavelength (X rays). At cptical frequencies it has definite advantages and is well described in the literature.

$$\nabla \times \mathbf{B} = \frac{1}{c}\frac{\partial \mathbf{E}}{\partial t} + \frac{4\pi}{c}\mathbf{J}$$

$$\nabla \times \mathbf{E} = -\frac{1}{c}\frac{\partial \mathbf{B}}{\partial t} \tag{3-2}$$

The volume averaged current density is usually expanded into a multipole series after the part due to convection currents has been split off.

$$\mathbf{J} = \mathbf{J}_{\text{cond}} + \frac{\partial \mathbf{P}}{\partial t} + c\nabla \times \mathbf{M} - \frac{\partial}{\partial t}(\nabla \cdot \mathbf{Q}) + \cdots \tag{3-3}$$

This expansion is not unique. If the unit volume element has over-all neutrality, the electric dipole moment is uniquely defined. **M** is not uniquely defined, unless $\partial \mathbf{P}/\partial t = 0$. The quadrupole moment is not uniquely defined, unless $\mathbf{P} = 0$. It is, nevertheless, a useful expansion if the volume can be divided into unit cells for which these moments can be calculated as successive approximations to **J**. They all may contain nonlinear parts besides the linear terms.

3-1 ENERGY CONSIDERATIONS

From Maxwell's equations one obtains in the usual way[2] the equation of energy conservation,[3]

$$\frac{c}{4\pi}\nabla \cdot \mathbf{E} \times \mathbf{B} + \frac{1}{4\pi}\mathbf{B} \cdot \frac{\partial \mathbf{B}}{\partial t} + \frac{1}{4\pi}\mathbf{E} \cdot \frac{\partial \mathbf{E}}{\partial t} + \mathbf{E} \cdot \mathbf{J} = 0 \tag{3-4}$$

Introduce the constitutive relations

$$\mathbf{H} = \mathbf{B} - 4\pi\mathbf{M} \qquad \text{and} \qquad \mathbf{D} = \mathbf{E} + 4\pi\mathbf{P}$$

This last definition implies $\nabla \cdot \mathbf{D} - 4\pi\nabla\nabla \cdot \mathbf{Q} = 4\pi\rho$. Substitution of Eq. (3-3) into Eq. (3-4) yields

$$\frac{c}{4\pi}\nabla \cdot \left(\mathbf{E} \times \mathbf{H} - \frac{4\pi}{c}\mathbf{E} \cdot \frac{\partial \mathbf{Q}}{\partial t}\right) + \frac{1}{4\pi}\mathbf{H} \cdot \frac{\partial \mathbf{B}}{\partial t} + \frac{1}{4\pi}\mathbf{E} \cdot \frac{\partial \mathbf{D}}{\partial t} + \nabla \mathbf{E} : \frac{\partial \mathbf{Q}}{\partial t} = 0 \tag{3-5}$$

Note that the presence of a time-varying quadrupole moment necessitates a redefinition of the Poynting vector, which is the term between brackets in Eq. (3-5). The rate of energy flow out of a volume element plus the rate of work done by the fields on a unit volume of the material system equals the decrease in stored energy density

$-(\partial/\partial t)(H^2/8\pi + E^2/8\pi)$. For a lossless medium the rate of work done per unit volume equals the rate at which the stored energy density U increases,

$$\frac{\partial U}{\partial t} = \mathbf{H}\cdot\frac{\partial \mathbf{M}}{\partial t} + \mathbf{E}\cdot\frac{\partial \mathbf{P}}{\partial t} + \nabla\mathbf{E} : \frac{\partial \mathbf{Q}}{\partial t} \qquad (3\text{-}6)$$

Consider another function, F, defined by

$$F = U - \mathbf{H}\cdot\mathbf{M} - \mathbf{E}\cdot\mathbf{P} - \nabla\mathbf{E}:\mathbf{Q} \qquad (3\text{-}7)$$

This energy is associated with the work done by the generators in order to establish the fields in the presence of the medium. It has a total differential

$$dF = -\mathbf{P}\cdot d\mathbf{E} - \mathbf{M}\cdot d\mathbf{H} - \mathbf{Q}:d\nabla\mathbf{E} \qquad (3\text{-}8)$$

If F is expressed as a function of the field strengths, the polarizations may be obtained by differentiation.

The discussion is so far quite general. Now the restriction is made to media with a small nonlinearity in the quasi-steady state condition. The fields will be expanded in a Fourier series with coefficients that are very slowly changing with time, e.g.,

$$E(\omega,t) = \frac{1}{2T}\int_{t-T}^{t+T} E(t)\exp(-i\omega t)\,dt = E^*(-\omega,t) \qquad (3\text{-}9)$$

The time T is chosen long enough so that $\omega T \gg 1$, but still short enough that $E(\omega,t)$ does not depend on T. This is not possible if the nonlinearity is strong. Fortunately the optical nonlinearities can always be considered as small perturbations and the description of the field in terms of a finite number of Fourier components is meaningful for a steady state. Considering the Fourier components of the polarizations and the fields at $\mathfrak{N}$ corresponding frequencies, all defined in the sense of Eq. (3-9), the time average increment is in the complex notation

$$\langle dF\rangle = -2\mathrm{Re}\sum_{i=1}^{\mathfrak{N}}[\mathbf{P}(\omega_i)\cdot d\mathbf{E}^*(-\omega_i) + \mathbf{M}(\omega_i)\cdot d\mathbf{H}^*(-\omega_i) + \mathbf{Q}(\omega_i):\nabla\mathbf{E}^*(-\omega_i)] \qquad (3\text{-}10)$$

The rapidly fluctuating terms in Eq. (3-8) are ignored.

To go from the steady state with Fourier amplitudes $E^*(-\omega_i)$, etc., to a new steady state with Fourier amplitudes $E^*(-\omega_i) + dE^*(-\omega_i)$, the generators have to do a certain amount of work in vacuum. In the presence of a medium with *prescribed* Fourier components of polarization, the amount of work to be done by the external generator is less by $\langle dF \rangle$. The Fourier components of polarization are obtained from this potential by differentiation

$$\mathbf{P}(\omega_i) = -\partial F/\partial \mathbf{E}^*(-\omega_i)$$

$$\mathbf{M}(\omega_i) = -\partial F/\partial \mathbf{H}^*(-\omega_i)$$

$$\mathbf{Q}(\omega_i) = -\partial F/\partial \nabla \mathbf{E}^*(-\omega_i)$$

For the linear case these relationships are familiar. Then F is a quadratic function in the field amplitudes, e.g., energy for the electric dipole case, $F^L = -\Sigma\, \mathbf{E}^*(-\omega_i) \cdot \chi(\omega_i) \cdot \mathbf{E}(\omega_i)$. If the change in free energy of the fields due to the presence of the lossless medium is given in terms of a power series in the field amplitudes, the nonlinear Fourier components in various order can be obtained. The nonlinear part of the time-averaged free energy can be related to the nonlinear susceptibilities. It was shown in Chapter 1 how the permutation relations of the nonlinear susceptibilities follow from the physical fact that the steady state reached is independent of the order in which the Fourier components reach their steady state value, provided the changes are slow in the sense of Eq. (3-9) so that the frequencies remain meaningful. In general F will be a joint power series in $\mathbf{E}$, $\mathbf{H}$, and $\nabla\mathbf{E}$. One may, e.g., derive electric polarization components proportional to a power of the magnetic field amplitude, etc. In a medium with inversion symmetry, the lowest order nonlinear terms in the potential F take the form

$$\begin{aligned}\left\langle F^{NL}_{symm} \right\rangle - \chi_1 \mathbf{EEH}^* - \chi_2 \mathbf{EE}^*\mathbf{H} - \chi_3 \mathbf{EE}\nabla\mathbf{E}^* - \chi_4 \mathbf{EE}^*\nabla\mathbf{E} \\ - \chi_5 \mathbf{EEEE}^* - \chi_6 \mathbf{EEE}^*\mathbf{E}^* + \text{complex conj}\end{aligned} \tag{3-11}$$

The sum of the negative frequencies of the starred quantities and the positive frequencies of the unstarred Fourier components equals zero.

In the presence of dispersion considerable care must be exercised in the calculation of the average internal energy U of the medium. During the slow change of the Fourier amplitude $E(\omega_i, t)$ to its final value, the field is not strictly monochromatic, but contains some neighboring frequencies $\omega_i + \eta$. Consider the linear dispersive medium

$$\mathbf{E}(t) = \int_{\omega_i-\delta}^{\omega_i+\delta} \mathbf{E}(\eta)\, e^{-i(\omega_i+\eta)t}\, d\eta$$

$$\frac{\partial \mathbf{E}(t)}{\partial t} = -i\omega_i \mathbf{E}(t) - i \int \eta \mathbf{E}(\eta)\, e^{-i(\omega_i+\eta)t}\, d\eta$$

$$\mathbf{P}(t) = \int \chi(\omega_i + \eta)\, \mathbf{E}(\eta) e^{-i(\omega_i+\eta)t}\, d\eta$$

Expand the susceptibility around the frequency ω_i and retain only the lowest power in η, since $E(\omega_i)$ is changed sufficiently slowly.

$$\mathbf{P}(t) = \boldsymbol{\chi}(\omega_i)\mathbf{E}(t) + (\partial\boldsymbol{\chi}/\partial\omega)_{\omega_i} \int \eta\, E(\eta) e^{-i(\omega_i+\eta)t}\, d\eta$$

$$= \boldsymbol{\chi}(\omega_i)\mathbf{E}(t) - \omega_i (\partial\boldsymbol{\chi}/\partial\omega)_{\omega_i} E(t) + i\frac{\partial \mathbf{E}}{\partial t} (\partial\boldsymbol{\chi}/\partial\omega)_{\omega_i}$$

Consider now the mean stored internal energy of the medium

$$\langle U \rangle = 2\, \mathrm{Re} \int_0^{E(\omega_i)} \mathbf{E}^* \cdot d\mathbf{P} = 2\, \mathrm{Re}\, \mathbf{E}^*(\omega_i) \cdot \mathbf{P}(\omega_i)$$

$$- 2\, \mathrm{Re} \int_0^{E(\omega_i)} \mathbf{P}(t) \cdot d\mathbf{E}^*(t)$$

$$= 2\, \mathrm{Re} \left[\mathbf{E}^*(\omega_i) \cdot \mathbf{P}(\omega_i) - \tfrac{1}{2}\mathbf{E}^*(\omega_i) \cdot \boldsymbol{\chi}(\omega_i) \cdot \mathbf{E}(\omega_i) \right.$$

$$\left. + \tfrac{1}{2}\omega_i \mathbf{E}^*(\omega_i) \cdot (\partial \boldsymbol{\chi}/\partial\omega)_{\omega_i} \cdot \mathbf{E}(\omega_i) \right]$$

Drop the subscript i on the frequency of the Fourier component. Except for a factor 4 due to our definition of the amplitudes, the familiar result is obtained,

$$\langle U \rangle = \mathbf{E}^*(\omega) \cdot \partial(\omega\boldsymbol{\chi})/\partial\omega \cdot \mathbf{E}(\omega) \qquad (3\text{-}12)$$

It is interesting to check that this expression indeed corresponds to the sum of kinetic and potential energy of driven harmonic oscillators, which may represent the linear dispersive medium.

In the presence of dispersion the transition from Eq. (3-8) to the time-averaged form Eq. (3-10) apparently needs some correction. Differentiation of the function F with respect to E_j yields

$$-\partial F/\partial E_j^* = \sum_i \{\chi_{ji}(\omega) + \omega(\partial\chi_{ji}/\partial\omega)\} E_i$$

If one also differentiates with respect to E_j, the following relation is obtained:

$$\chi_{ij}(\omega) + \omega\partial\chi_{ij}/\partial\omega = \chi_{ji}(\omega) + \omega\partial\chi_{ji}/\partial\omega$$

From this the symmetry of the linear susceptibility tensor still follows, $\chi_{ij}(\omega) = \chi_{ji}(\omega)$.

These considerations may be extended to nonlinear media in the presence of dispersion. If the field at ω_1 is slowly increased in amplitude while the amplitudes at ω_2 and $\omega_3 = \omega_1 + \omega_2$ are held constant, the amount of work done on the medium due to the nonlinearity is

$$2\,\mathrm{Re}\Big[\int_0^{E(\omega_1)} \mathbf{E}_3^* \cdot d\mathbf{P}^{NL}(\omega_1 + \omega_2, t) + \int_0^{E(\omega_1)} \mathbf{E}_2 \cdot d\mathbf{P}^{*NL}(\omega_1 - \omega_3, t)\Big]$$

The variation of the amplitude $E(\omega_1)$ is sufficiently slow that it never has Fourier amplitudes at ω_2 and ω_3. The integrands E_3^* and E_2 are independent of $E(\omega_1, t)$ and the increase in the time-averaged internal energy due to the nonlinearity is

$$\langle U^{NL}\rangle = 2\,\mathrm{Re}\,[\mathbf{E}_3^* \cdot \mathbf{P}^{NL}(\omega_1 + \omega_2) + \mathbf{E}_2 \cdot \mathbf{P}^{*NL}(\omega_1 - \omega_3)]$$

The corresponding change in F is

$$\langle F^{NL}\rangle = -2\,\mathrm{Re}\,\mathbf{E}_1 \cdot \mathbf{P}^{*NL}(\omega_2 - \omega_3)$$

$$= -2\,\mathrm{Re}\,[\chi^{NL} E_1 E_2 E_3^*]$$

When all frequencies are different, the presence of dispersion provides no further correction. When two frequencies are equal, corrections involving the derivatives of the nonlinear susceptibility with respect to this frequency occur, in a manner analogous to the linear case. The permutation symmetry relations retain their validity even then. The most important application of the permutation symmetry relation occurs when the dispersion in the immediate vicinity of the frequencies ω_1, ω_2, ω_3, etc. is very small, but there are important dispersion and absorption regions between these frequencies. In that case the rather subtle energy considerations in the presence of dispersion are not important.

In regions of strong dispersion there is always a concomitant loss mechanism. For a lossy medium it will sometimes be possible to derive the rate of absorption or entropy production in the steady state from a dissipation function. Pershan[3] has discussed this and pointed out the relationship to the Onsager relations in the linear case. Since the most useful symmetry property for the part of the nonlinear susceptibility corresponding to loss has already been derived in Chapter 2 in the microscopic theory, no further discussion will be given here. Several applications of the free energy terms in Eq. (3-11) are discussed in Chapter 5.

3-2 LOCAL FIELDS IN OPTICALLY DENSE MEDIA

The atomic models used in Chapters 1 and 2 expressed the dipole moment of a molecule or molecular group in terms of the electric field acting on it. Only in dilute gases can this acting or local field be identified with the macroscopic field appearing in Maxwell's equations. In dense madia the influence of neighboring dipoles must be taken into account. If the polarizable unit samples with equal weight all volume elements in a unit cell, the average acting field is equal to the macroscopic field. This is essentially the case for valence electrons in semiconductors and metals. For well-localized ions or molecular groups the difference between the acting local field and the macroscopic field may be calculated by a method due to Lorentz. Consider for simplicity only the case of one ion at a site with cubic symmetry. The acting field is

$$\mathbf{E}_{\text{loc}} = \mathbf{E} + \frac{4\pi}{3}\mathbf{P} = \mathbf{E} + \frac{4\pi}{3}\mathbf{P}^{\text{L}} + \frac{4\pi}{3}\mathbf{P}^{\text{NL}} \tag{3-13}$$

The polarization consists of both a linear and a nonlinear part. Denote the linear polarizability of the ion by α, its nonlinear polarizability by $\chi_{ijk}^{\text{at}}(\omega_3 = \omega_1 + \omega_2)$. If there are N ions per cubic centimeter, the linear polarization is

$$\mathbf{P}^{\text{L}} = N\alpha\mathbf{E}_{\text{loc}} \tag{3-14}$$

The nonlinear polarization is

$$\mathbf{P}^{\text{NL}} = N\chi^{\text{at}}\mathbf{E}_{1,\text{loc}}\mathbf{E}_{2,\text{loc}} \tag{3-15}$$

The displacement vector is

$$\mathbf{D} = \mathbf{E} + 4\pi\mathbf{P}^{\text{L}} + 4\pi\mathbf{P}^{\text{NL}} \tag{3-16}$$

Combination of Eq. (3-13) and (3-14) gives

$$\mathbf{P}^{\text{L}} = \frac{N\alpha}{1 - 4\pi N\alpha/3}\left(\mathbf{E} + 4\frac{\pi}{3}\mathbf{P}^{\text{NL}}\right) = \frac{\epsilon - 1}{4\pi}\left(\mathbf{E} + \frac{4\pi}{3}\mathbf{P}^{\text{NL}}\right)$$

Substitution of this expression into Eq. (3-16) yields

$$\mathbf{D} = \epsilon \mathbf{E} + 4\pi \frac{\epsilon + 2}{3} \mathbf{P}^{NL} = \epsilon \mathbf{E} + 4\pi \mathbf{P}^{NLS} \tag{3-17}$$

In a dense medium the effective nonlinear source polarization to be used in Maxwell's equations is $\frac{1}{3}(\epsilon + 2)$ times the true induced nonlinear dipole moment multiplied by the number of ions per cubic centimeter. Combination of Eqs. (3-15), (3-16), and (3-17) finally gives the relationship between the nonlinear source term and the nonlinear polarizability of the isolated system

$$\begin{aligned} \mathbf{P}^{NLS}(\omega_3) &= \frac{\epsilon(\omega_1) + 2}{3} \frac{\epsilon(\omega_2) + 2}{3} \frac{\epsilon(\omega_3) + 2}{3} N\chi^{at}(\omega_3 = \omega_1 + \omega_2) \\ &\quad \times \mathbf{E}(\omega_1)\mathbf{E}(\omega_2) \qquad (3\text{-}18) \\ &= \chi^{mac}(\omega_3 = \omega_1 + \omega_2)\mathbf{E}(\omega_1)\mathbf{E}(\omega_2) \end{aligned}$$

Because the correction factor is symmetric in the three frequencies, the permutation symmetry relations remain valid for the macroscopic nonlinear susceptibility, as they were for χ^{at}. The argument given here applies perhaps only to a substance like CuCl, which is rather ionic with ions at sites with $\bar{4}3$ m symmetry. This tetrahedral symmetry is cubic but lacks an inversion center. The correction factors are of course very significant for large dielectric constants. It is, however, assumed that the valence electron is localized around a lattice point with cubic symmetry. This is certainly *not* the case for valence electrons in semiconductors like GaAs. Then a better approximation is to put the local field correction factor equal to unity. It is doubtful whether in CuCl the full correction of Eq. (3-18) will apply.

Eq. (3-17) remains valid for higher order nonlinearities. For third-harmonic generation in alkali halide crystals it is appropriate to apply the correction factor $(\epsilon(3\omega) + 2)(\epsilon(\omega) + 2)^3/81$. A similar correction factor is appropriate in fluids. For the Raman effect in molecular liquids without association one would apply a correction factor $(\epsilon(\omega_L) + 2)^2(\epsilon(\omega_S) + 2)(\epsilon(\omega_a) + 2)/81$ to $\chi^{at}(\omega_a = 2\omega_L - \omega_S)$.

Similar arguments can be given for systems localized at sites of lower symmetry and for situations with several different systems in the unit cell. The Lorentz correction factors become tensorial structures. The permutation symmetry relations remain valid, as could be expected from the general energy considerations of the preceding section. Details are given in the appendix to the paper in Appendix I.

The correct local field corrections in complex geometrics, such as occur, e.g., in a KH_2PO_4 crystal, are difficult to determine. Since the corrections are by no means small, they represent a large source of uncertainty.

3-3 COUPLED WAVE EQUATIONS IN NONLINEAR MEDIA

With the determination, in principle, of the effective macroscopic nonlinear polarization, one may return to Maxwell's equations. It is useful to write these equations for each Fourier component with the nonlinear part split off. In this manner the nonlinear source (NLS) terms are clearly exhibited. Assume for simplicity a nonmagnetic medium, $\mu^L = 1$.

$$\begin{aligned} \nabla \times \mathbf{E}(\omega_i) &= +i(\omega_i/c)\mathbf{H}(\omega_i) \\ \nabla \times \mathbf{H}(\omega_i) &= -i(\omega_i/c)\boldsymbol{\epsilon}(\omega_i)\cdot\mathbf{E}(\omega_i) + (4\pi/c)\mathbf{J}^{NLS}(\omega_i) \end{aligned} \tag{3-19}$$

Combination of these eauations leads to the nonlinear wave equation

$$\nabla \times \nabla \times \mathbf{E}(\omega_i) - (\omega_i/c)^2\boldsymbol{\epsilon}(\omega_i)\cdot\mathbf{E}(\omega_i) = 4\pi i(\omega_i/c^2)\mathbf{J}^{NLS}(\omega_i) \tag{3-20}$$

If $J^{NLS}(\omega_i)$ is expressed in terms of other Fourier components of the field, a set of coupled wave equations results. The complexity of the solutions increases of course very rapidly with the number of Fourier components that are retained. It is very important to truncate the number of components by physical arguments to obtain approximate solutions.

In terms of the electric dipole case, Maxwell's equations can be written for each Fourier component as

$$\begin{aligned} \nabla \times \mathbf{E}(\omega_i) &= +i(\omega_i/c)\mathbf{H}(\omega_i) \\ \nabla \times \mathbf{H}(\omega_i) &= -i(\omega_i/c)\boldsymbol{\epsilon}(\omega_i)\cdot\mathbf{E}(\omega_i) - 4\pi i(\omega_i/c)\mathbf{P}^{NLS}(\omega_i) \end{aligned} \tag{3-21}$$

The wave equation becomes

$$\nabla \times \nabla \times \mathbf{E}(\omega_i) - (\omega_i/c)^2\boldsymbol{\epsilon}(\omega_i)\mathbf{E}(\omega_i) = 4\pi(\omega_i/c)^2\mathbf{P}^{NLS}(\omega_i) \tag{3-22}$$

Consider explicitly the lowest order nonlinearity with three waves at frequencies $\omega_3 = \omega_1 + \omega_2$, ω_1, and ω_2. One has the three complex nonlinear coupled vector wave equations,

$$\nabla \times \nabla \times \mathbf{E}(\omega_1) - (\omega_1^2/c^2)\boldsymbol{\epsilon}(\omega_1) \cdot \mathbf{E}_1(\omega_1)$$
$$= 4\pi(\omega_1^2/c^2)\boldsymbol{\chi}(\omega_1 = \omega_3 - \omega_2)\mathbf{E}(\omega_3)\mathbf{E}_2^*(-\omega_2)$$
$$\nabla \times \nabla \times \mathbf{E}(\omega_2) - (\omega_2^2/c^2)\boldsymbol{\epsilon}(\omega_2) \cdot \mathbf{E}_2(\omega_2) \tag{3-23}$$
$$= 4\pi(\omega_2^2/c^2)\boldsymbol{\chi}(+\omega_2 = \omega_3 - \omega_1)\mathbf{E}_3(\omega_3)\mathbf{E}_1^*(-\omega_1)$$
$$\nabla \times \nabla \times \mathbf{E}(\omega_3) - (\omega_3^2/c^2)\boldsymbol{\epsilon}(\omega_3) \cdot \mathbf{E}_3(\omega_3)$$
$$= 4\pi(\omega_3^2/c^2)\boldsymbol{\chi}(\omega_3 = \omega_1 + \omega_2)\mathbf{E}_1(\omega_1)\mathbf{E}_2(\omega_2)$$

These equations are still useful and valid in lossy media. In this case both $\boldsymbol{\epsilon}$ and $\boldsymbol{\chi}$ are complex quantities. In the case of small loss $\epsilon(\omega_1)$, $\epsilon(\omega_2)$ and $\epsilon(\omega_3)$ are taken as complex quantities, but one ignores the imaginary part in the nonlinear coupling. The nonlinear susceptibilities then still obey the permutation symmetry relations in Eq. (1-28). When there is strong damping near only one frequency, say ω_1, the complex symmetry relation applies,

$$\chi_{ijk}(\omega_1 = \omega_3 - \omega_2) = \chi^*_{kji}(\omega_2 = \omega_3 - \omega_1) = \chi_{jik}(\omega_3 = \omega_1 + \omega_2)$$

In the most general case no simple symmetry relation exists.

Wave equations of the type (3-23) have to be solved, subject to the appropriate boundary conditions. The tangential components of **E** and **H** must be continuous at the boundary, for each Fourier component separately, because the general boundary conditions must be met at all times. In a similar manner the normal components of **D** and **B** of each Fourier component must be continuous.

The problem can not be limited, in principle, to three Fourier components, because the nonlinearity will create further combination frequencies $\pm\ell_1\omega_1 \pm \ell_2\omega_2 \pm \ell_3\omega_3$. It is hoped that in practice one may exclude most sets of integers ℓ_1, ℓ_2, and ℓ_3 because the amplitudes of the corresponding waves will be extremely small. At least one situation will be encountered, in the Raman effect, where this hope is not fulfilled and the set of coupled waves has to be extended to a large number of waves. The hope for an analytic approximate solution must then usually be abandoned. In the next chapter a variety of situations of physical interest will be discussed. Physically meaningful approximations allow for solutions of varying complexity and accuracy. The weak nonlinearity of the optical properties is of course the fundamental reason that such solutions can be obtained.

3-4 A PARTICULAR SOLUTION FOR ARBITRARY NONLINEAR RESPONSE

The only solution for a problem in optics with arbitrary nonlinearity has been given by Broer.[4] He considered an electromagnetic wave in vacuum with an arbitrary time dependence **E**(t) in its amplitude, incident normally on the boundary of a nonlinear medium with an arbitrary nonlinear response without dispersion. Broer's solution for the reflected wave will be reproduced here to exhibit the entirely different nature of the problem, when the assumption of small nonlinearity is not valid.

A nonmagnetic medium (B = H) fills a half space, $z > 0$. It has an arbitrary functional relationship between D or P^{NLS} and E. Maxwell's equations for one linear polarization taken in the y-direction in the medium are,

$$\frac{\partial E_y}{\partial z} - \frac{1}{c}\frac{\partial B_x}{\partial t} = 0 \qquad \frac{\partial B_x}{\partial z} = \frac{1}{c}\frac{\partial D_y}{\partial t} \tag{3-24}$$

Henceforth the subscripts x and y will be omitted. Define a function $v^{-2}(E) = (dD/dE)\cdot c^{-2}$. The equations can be written as

$$\frac{\partial B}{\partial t} = c\frac{\partial E}{\partial z} \qquad \frac{\partial B}{\partial z} = v^{-2}(E)\, c\frac{\partial E}{\partial t} \tag{3-25}$$

A solution can be written in the differential form,

$$\begin{aligned} dE &= v/c\, dB \quad \text{for} \quad dz/dt = v \\ dE &= -v/c\, dB \quad \text{for} \quad dz/dt = -v \end{aligned} \tag{3-26}$$

There exist solutions in the nonlinear medium in which the wave travels away from the boundary. If the time variation at the boundary is prescribed $E_y(z = 0) = G(t)$, then the solution in the medium is

$$E = G\left\{t - \frac{z}{v(E)}\right\} \tag{3-27}$$

The function G(t) is of course in turn determined by the incident and reflected waves in the vacuum, $z < 0$. The former is assumed to be given $f_{inc}(t - z/c)$. If the reflected wave is denoted by $f_R(t + z/c)$, the vacuum fields are

$$\begin{aligned} E_{vac}(z < 0) &= f_{inc}(t - z/c) + f_R(t + z/c) \\ B_{vac} &= -f_{inc}(t - z/c) + f_R(t + z/c) \end{aligned} \tag{3-28}$$

The latter solution has to be matched to the magnetic field inside the medium at $z = 0$. Integration of Eq. (3-26) gives

$$B(z = 0,t) = -\int_0^{G(t)} (c/v)\, dE = -\int_0^{G(t)} \left(\frac{dD}{dE}\right)^{1/2} dE \qquad (3\text{-}29)$$

Matching the tangential components at $z = 0$, one finds the implicit relation between the reflected and the incident wave,

$$f_i - f_r = \int_0^{f_i + f_R} \left(\frac{dD}{dE}\right)^{1/2} dE \qquad (3\text{-}30)$$

In the following chapter the reflected wave from a nonlinear medium will be obtained as a Fourier series containing the harmonics and combination tones of the incident quasi-monochromatic waves. That solution is an excellent and rapidly converging approximation to the exact solution (3-30) because the nonlinearities are small. The approximate solution is better adapted to the experimental situation, where a spectrographic Fourier analysis of the reflected light is performed. It is also readily extended to other than normal incidence, anisotropic nonlinear media, etc. The exact solution (3-30) should, however, serve as a reminder of the basic approximation involved in the consideration of a limited number of Fourier components.

REFERENCES

1. L. Rosenfeld, *Theory of Electrons*, North-Holland Publishing Co., Amsterdam, 1949.
2. M. Born and E. Wolf, *Principles of Optics*, Pergamon Press, London, 1959, Chap. 1.
3. P. S. Pershan, *Phys. Rev.*, **130**, 919 (1963).
4. L. J. F. Broer, *Phys. Letters*, **4**, 65 (1963).

4

WAVE PROPAGATION IN NONLINEAR MEDIA

4-1 PARAMETRIC GENERATION AND BOUNDARY CONDITIONS

In this section the birth of a wave at a new frequency will be analyzed. Consider a nonlinear medium traversed by a number of waves with frequencies $\omega_1 \ldots, \omega_i \ldots,$ wave vectors $\mathbf{k}_1 \ldots, \mathbf{k}_i \ldots,$ and polarization vectors $\hat{a} \ldots, \hat{a}_i \ldots.$ These waves will generate a polarization P^{NLS} at the combination frequency $\omega_s = \Sigma_i \ell_i \omega_i$. This polarization will generate radiation at ω_s. At first the amplitude of the new wave at ω_s will be so small that its reaction on the original waves may be ignored. The fields at the original frequencies are regarded as fixed parameters. In this parametric approximation only one wave equation at the time need be considered, or rather two for two directions of polarization.

The nonlinear source distribution is given by

$$\mathbf{P}^{NLS}(\omega_s, \mathbf{r}) = \mathbf{P}_0^{NLS} \exp(i\mathbf{k}_s \cdot \mathbf{r} - i\omega_s t) \tag{4-1}$$

with

$$\omega_s = \sum_i \ell_i \omega_i \qquad \mathbf{k}_s = \sum_i \ell_i \mathbf{k}_i$$

$$P_0^{NLS} = \chi^{NL} \prod_i (\hat{\mathbf{a}}_i E_i)^{|\ell_i|}$$

where χ^{NL} is a nonlinear tensor of rank $\Sigma_i |\ell_i| + 1$. If ℓ_i is a negative integer $(\hat{a}_i E_i)$ has to be replaced by its complex conjugate. The product sign is merely a shorthand notation for repeated vectors. The important practical cases are of course only those with a small set of frequencies and small integers ℓ_i. Second-harmonic generation will be described by $i = 1$ and $\ell = 2$, etc.

The wave equation at the new frequency component ω_s takes the form,

$$\nabla \times \nabla \times \mathbf{E}(\omega_s) - \frac{\epsilon(\omega_s)\omega_s^2}{c^2}\mathbf{E}_s = \frac{4\pi\omega_s^2}{c^2}\mathbf{P}_0^{NLS} \exp(i\mathbf{k}_s \cdot \mathbf{r} - i\omega_s t) \tag{4-2}$$

The solution consists in general of one particular solution of the inhomogeneous equation with right-hand side and the general solution of the homogeneous equation with the right-hand side put equal to zero. That solution consists of a linear combination of all the plane waves or any other orthonormal set that can propagate in the linear medium at the frequency ω_s. Which linear combination should be considered is determined by the boundary conditions.[1]

Assume that the nonlinear medium has a plane boundary at $z = 0$. The incident light beams are assumed to be plane waves of infinite extent. The corrections necessary for finite cross section will be made in a later chapter. The assumed geometry is appropriate for the majority of nonlinear optical experiments. The direction of the incident beams immediately determines the tangential component, k_x, k_y, of the nonlinear source. On reflection and refraction the transverse components of momentum are conserved because the boundary conditions must be satisfied everywhere in the plane $z = 0$. One therefore finds immediately,

$$k_x^s = \sum \ell_i k_{ix} \qquad k_y^s = \sum \ell_i k_{iy}$$

In these equations k_{ix} and k_{iy} stand for the tangential components of the incident wave at ω_i, which are equal to the tangential components of this wave after refraction. After they have entered the medium, they create the nonlinear polarization. Only those plane waves at ω_s which have the same tangential wave vector components are acceptable and may satisfy the boundary conditions. This determines the direction of the waves, acceptable as homogeneous solutions. The direction of the inhomogeneous solution or polarization wave is determined by the normal component

$$k_z^s = \sum \ell_i k_{iz}^T$$

In this formula it is important to take the normal components of the waves transmitted through the medium after refraction.

Many salient features of parametric generation can best be illustrated by consideration of a medium with a scalar dielectric constant

ϵ, i.e., a cubic crystal or an isotropic fluid. Choose, without loss of generality, the y-axis in such a direction that $k_y^S = 0$. As in the case of linear refraction, a distinction may be made between the wave at ω_S polarized normal to the plane of transmission, which is created by the component $P_y^{NLS} = P_\perp^{NLS}$, and a wave whose electric field vector lies in the xz plane, created by the source component $P_\parallel^{NLS}$ lying in that plane.

The general solution of Eq. (4-2), consistent with these boundary conditions, is immediately seen to be

$$E_\perp = A_\perp^T e^{+i\mathbf{k}^T\cdot\mathbf{r}} + \frac{4\pi(\omega_S/c)^2 P_\perp^{NLS}}{|k^S|^2 - |k^T|^2} e^{ik_S\cdot\mathbf{r}} \tag{4-3}$$

where $\mathbf{k}^T$ is the wave vector of the transmitted homogeneous solution. It has a magnitude $|k^T| = \omega_S c^{-1}\epsilon^{1/2}$ and a tangential component $k_x^T = k_x^S$. The last term represents the particular solution of the inhomogenous equation. The amplitude of the homogeneous solution $A_\perp^T$ must be determined from the requirement that the tangential components of E and H are continuous. These two conditions require also the presence of a reflected wave at ω_S, emanating from the boundary back into the linear medium,

$$E_\perp^R = A_\perp^R e^{i\mathbf{k}^R\cdot\mathbf{r}} \tag{4-4}$$

where

$$k_x^R = k_x^T = k_x^S$$

and

$$|\mathbf{k}^R| = \omega_S c^{-1}\epsilon_R^{1/2}$$

The fact that there must be a field at ω_S in the linear medium can be understood as follows. Maxwell's Eq. (3-2) gives a magnetic field proportional to $\partial P^{NLS}/\partial t$. Since there is a magnetic field in the nonlinear medium for $z > 0$, there must also be a magnetic field at ω_S in the linear medium for $z < 0$. The subscript "s" on ω_S will be dropped.

The continuity condition at $z = 0$ for E_y is simply

$$A_\perp^T + \frac{4\pi(\omega/c)^2 P_\perp^{NLS}}{|k^S|^2 - |k^T|^2} = A_\perp^R \tag{4-5}$$

The continuity condition at $z = 0$ for H_x is,

$$(\omega/c)\epsilon_T^{1/2} A_\perp^T \cos\theta_T + \frac{|k^S|\, 4\pi(\omega/c)^2 P_\perp^{NLS}}{|k^S|^2 - |k^T|^2} \cos\theta_S$$
$$= -(\omega/c)\epsilon_R^{1/2} A_\perp^R \cos\theta_R \qquad (4\text{-}6)$$

Here the angle of reflection θ_R, the angle of transmission θ_T for the homogeneous wave, and θ_S for the polarization source have been introduced. These angles are shown in Figure 4-1 and are analytically defined by,

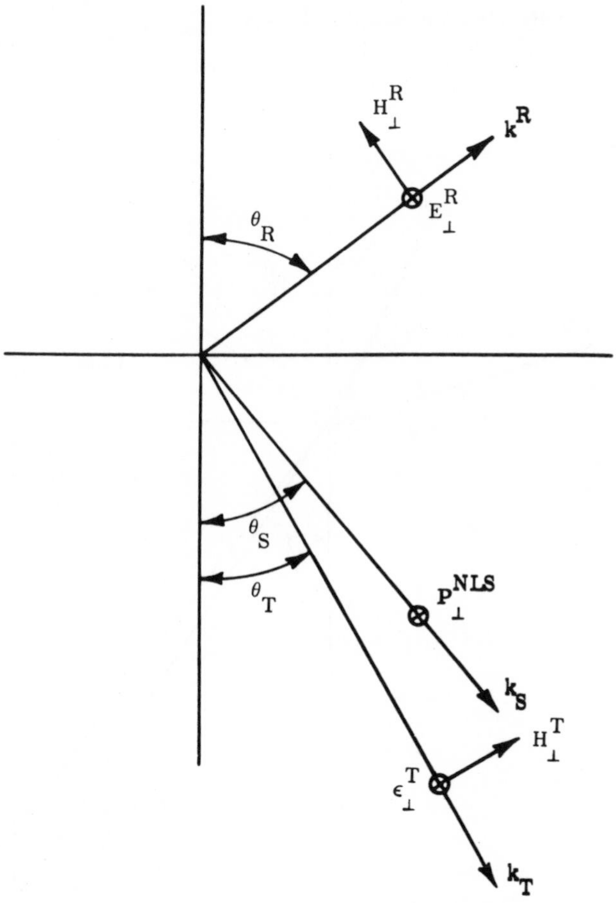

Figure 4-1. The angles for the nonlinear polarization and the transmitted and reflected waves generated by this polarization, for the case that the electric vectors are normal to the plane of reflection.

$$\epsilon_T^{1/2}\omega c^{-1}\sin\theta_T = \epsilon_R^{1/2}\omega c^{-1}\sin\theta_R = |k_S|\ \sin\theta_S = \sum_i \ell_i k_{ix} \tag{4-7}$$

$$0 = \sum_i \ell_i k_{iy}$$

The directions of all waves are determined by the angles of incidence and the planes of incidence of the waves at the composing frequencies. This is illustrated in Figure 4-2. Eqs. (4-5) and (4-6) can be solved for $A_\perp^T$ and $A_\perp^R$. Introduction of $\epsilon_S^{1/2} = c\,|k_S|\,\omega_S^{-1}$ yields the amplitude of the reflected wave in the form,

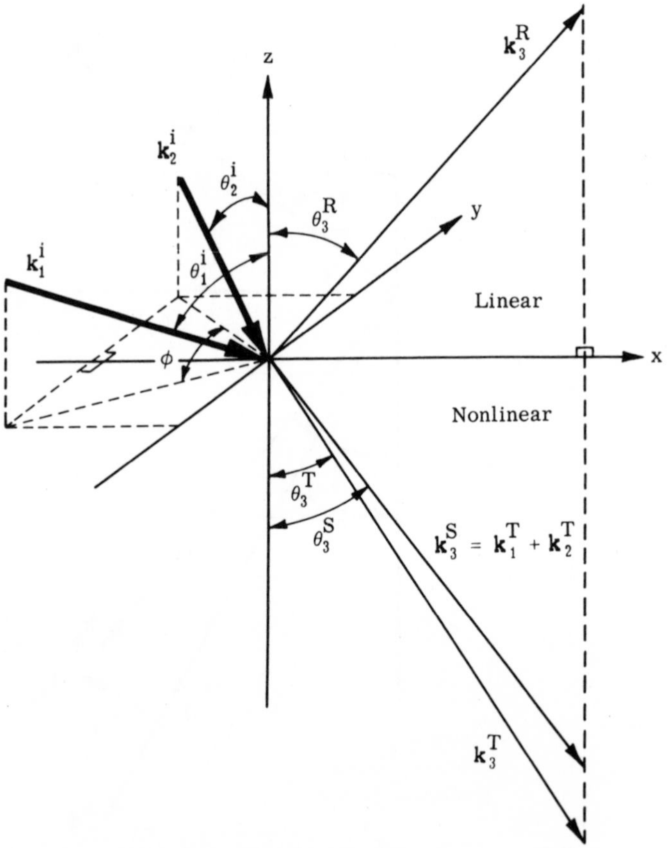

Figure 4-2. Geometrical relationship between the wave vectors of two plane waves, incident on the plane boundary of a nonlinear medium, and the reflected and transmitted waves at the sum frequency.

$$A_{\perp}^{R} = -4\pi P_{\perp}^{NLS} [(\epsilon_T^{1/2} \cos\theta_T + \epsilon_R^{1/2} \cos\theta_R) \times (\epsilon_T^{1/2} \cos\theta_T + \epsilon_S^{1/2} \cos\theta_S)]^{-1} \quad (4\text{-}8)$$

The reflected wave at the second harmonic frequency was first observed by Ducuing and Bloembergen.[2] They used a crystal of GaAs, which has the $\bar{4}3m$ symmetry so that the theory given here applies. In Figure 4-3 an experimental arrangement is shown where two rays at the fundamental beam strike the nonlinear crystal with different angles of incidence, $\sin\theta_1 = k_{1X}/k$, $\sin\theta_2 = k_{2X}/k$. They produce three polarization waves at the second harmonic frequency with three corresponding reflected rays, $\sin\theta_{11}^R(2\omega) = 2k_{1X}(\omega)/k(2\omega)$ $\sin\theta_{12}^R(2\omega) = \{(k_{1X}(\omega) + k_{2X}(\omega)\}/k(2\omega)$ and $\sin\theta_{22}^R = 2k_{2X}(\omega)/k(2\omega)$.

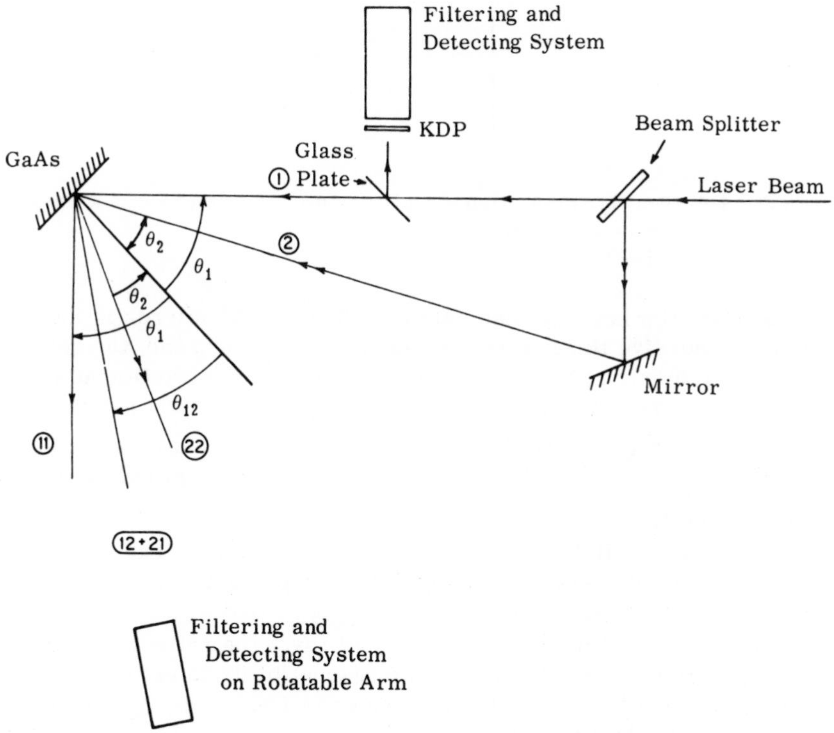

Figure 4-3. Experimental arrangement to detect second-harmonic generation in reflection. Two beams at ω, obtained by splitting a laser beam, create three beams at 2ω in reflection from a GaAs crystal.

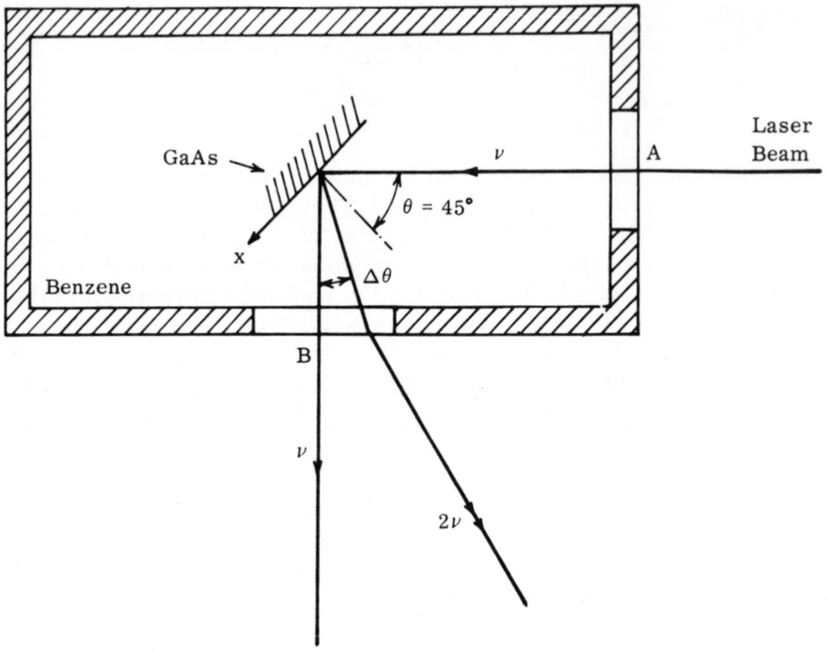

Figure 4-4. The reflection from a nonlinear mirror immersed in a linear dispersive medium.

This reflection law has been verified. If the GaAs mirror is inserted into a dispersive linear medium, such as liquid benzene, the reflected second harmonic ray does not appear in the same direction as the reflected fundamental ray, as shown in Figure 4-4. One has $\sin \theta^{i}(\omega) = \sin \theta^{R}(\omega) = k_{1x}(\omega)/|k(\omega)|$, and $\sin \theta^{R}(2\omega) = 2k_{1x}(\omega)/|k(2\omega)| = \{\epsilon_R^{1/2}(\omega)/\epsilon_R^{1/2}(2\omega)\} \sin \theta^{R}(\omega)$. This has also been verified experimentally by Ducuing.[3]

The GaAs crystal is strongly absorbing at both the ruby laser frequency and its second harmonic. It is important to note that all equations remain valid for complex ϵ_R, ϵ_T, and ϵ_S. Because of the strong absorption of the transmitted rays, the reflected intensity is easier to observe. The nonlinear susceptibility is also larger due to enhancement by resonant denominators. The effective layer that contributes to the harmonic ray intensity is contained in the smaller of the absorption depths at ω and 2ω. This is expressed in Eq. (4-8) by the fact that the denominator contains ϵ_S and ϵ_T. The observed intensity is of course proportional to the absolute square of Eq. (4-8) and therefore proportional to $|\chi^{NL}|^2$. The determination of the nonlinear susceptibility from the reflected intensity as well as its polarization properties will be discussed further in the next chapter. The expression for the reflected amplitude $E_{\parallel}^{R}$ polarized in the plane of reflection is given by Eq. (4-12) of Appendix 2. It is due to the component of the nonlinear source in that plane, $P_{\parallel}^{NLS}$. There is an analogue of Brewster's angle for harmonic reflection. For a certain direction of $P_{\parallel}^{NLS}$, $E_{\parallel}^{R}$ vanishes. For details Appendix 2 should be consulted.

The transmitted wave with perpendicular polarization is obtained from Eqs. (4-3), (4-5), and (4-8). With the use of the relations (4-7), the transmitted homogeneous amplitude can be brought into the form,

$$A_{\perp}^{T} = \frac{4\pi P^{\perp NLS}}{\epsilon_T - \epsilon_S} \times \frac{\epsilon_S^{1/2}\cos\theta_S + \epsilon_R^{1/2}\cos\theta_R}{\epsilon_T^{1/2}\cos\theta_T + \epsilon_R^{1/2}\cos\theta_R} \tag{4-9}$$

It follows from Figure 4-1 that

$$\mathbf{k}_S - \mathbf{k}_T = \omega c^{-1}\left[\epsilon_S^{1/2}\cos\theta_S - \epsilon_T^{1/2}\cos\theta_T\right] z$$

The transmitted wave may be written as a single wave with a propagation vector $\mathbf{k}^T$, but with an amplitude that depends on the distance z from the boundary. One derives from Eqs. (4-3) and (4-5) the transmitted wave

$$E_{\perp}^{T} = \left[A_{\perp}^{R} + 4\pi P_{\perp}^{NLS}\frac{\exp[i\omega c^{-1}(\epsilon_S^{1/2}\cos\theta_S - \epsilon_T^{1/2}\cos\theta_T)z] - 1}{\epsilon_S - \epsilon_T}\right] e^{i\mathbf{k}_T\cdot\mathbf{r} - i\omega t} \tag{4-10}$$

For values of z which satisfy the condition

$$\omega c^{-1} z(\epsilon_S^{1/2}\cos\theta_S - \epsilon_T^{1/2}\cos\theta_T) \ll 1 \tag{4-11}$$

the wave behaves as

$$E_{\perp}^{T} = \Big[A_{\perp}^{R} + 4\pi i P_{\perp}^{NLS}(\omega c^{-1} z) \times (\epsilon_S^{1/2}\cos\theta_S + \epsilon_T^{1/2}\cos\theta_T)^{-1}\Big] e^{i\mathbf{k}_T\cdot\mathbf{r} - i\omega t} \tag{4-12}$$

The wave starts off with an amplitude equal to that of the reflected wave and the component which is 90° out-of-phase with P^{NLS} grows proportional to the distance from the boundary. The wave does negative work on P^{NLS}. The intensity increases as z^2, as long as P^{NLS} in all unit cells retains its phase relation with respect to the wave. This will be the case for arbitrary distances z if the phase velocities normal to the boundary are matched, i.e.,

$$\epsilon_S^{1/2} \cos\theta_S = \epsilon_T^{1/2} \cos\theta_T$$

The phase velocities parallel to the boundary are necessarily matched. With the definition of ϵ_S, this is only possible if $\epsilon_S = \epsilon_T$, $\theta_S = \theta_T$, and $\mathbf{k}_S = \mathbf{k}_T$. Due to natural color dispersion this relationship will not be satisfied in cubic crystals for harmonic generation. In noncubic crystals phase matching of the harmonic with the fundamental is possible by compensation of the color dispersion by the phenomenon of double refraction. In that case the intensity of the harmonic wave appears to increase beyond all limits as z is increased. As soon as the intensity has built up to an appreciable fraction of the incident wave, the parametric approximation ceases to be valid. This question of reaction of the harmonic wave back on the fundamental will be discussed in section 4-2.

When the phase-matching condition is not satisfied, the intensity of the parametrically generated wave will vary sinusoidally with the distance z. The distance between two successive maxima is given by $\omega c^{-1}(\epsilon_S^{1/2}\cos\theta_S - \epsilon_T^{1/2}\cos\theta_T)z = 2\pi$. In general the transmitted wave is an inhomogeneous wave. The planes of constant amplitude (constant z) do not coincide with the planes of constant phase. An exception is the case of propagation normal to the boundary. In this case the intensity in the wave is determined by Eq. (4-10), if the small term $E_\perp^R$ is omitted:

$$|E_\perp^T|^2 = 16\pi^2 \, |P_\perp^{NLS}|^2 \, \frac{4\sin^2 \frac{1}{2}\omega c^{-1}(\epsilon_S^{1/2} - \epsilon_T^{1/2})z}{(\epsilon_S - \epsilon_T)^2} \tag{4-13}$$

Comparison of this equation with Eq. (4-8) shows that the maximum intensity in the transmitted wave is larger by a factor of the order $4\,\epsilon_T^2/(\epsilon_T - \epsilon_S)^2$ than the reflected intensity. For usual mismatch due to dispersion this expression is about $10^4 - 10^5$. The omission of $E_\perp^R$ in the transmitted wave is justified.

The periodic variation of the second harmonic intensity as the optical thickness of the nonlinear sample varies was first demonstrated by Maker and Terhune. Their result was shown in Figure

1-2. From such curves the intensity of the maxima is a measure for the nonlinear susceptibility, while the distance between maxima gives the coherence length or inverse phase mismatch $(\epsilon_S^{1/2} - \epsilon_T^{1/2})^{-1}$. Introducing indices of refraction n_s and n_t and a vacuum wave length λ_0, one may define

$$1_{coh} = 2\pi(c/\omega)(\epsilon_S^{1/2} - \epsilon_T^{1/2})^{-1} = \lambda_0(n_S - n_T)^{-1} \tag{4-14}$$

It is the distance over which the polarization wave and the homogeneous wave keep in step. The reflected intensity is much weaker because only the atoms in a layer about $\frac{1}{2}\lambda/n_T$ thick contribute in phase to this wave.

In the experimental arrangements the nonlinear material is usually a plane-parallel slab. Only if the back side were made nonreflecting for both the fundamental and harmonic waves, could the present analytical results be compared directly with experiment. A complete analysis of the plane-parallel slab is given in Appendix 2. It involves matching of boundary conditions simultaneously on back and front boundary with backward and forward traveling waves. The correction for the transmitted harmonic wave is usually small for small Fresnel reflectivities. The small amount of reflected intensity at the back surface is, however, sufficient to given an entirely different value for $E_\perp^R$. Therefore the nonlinear reflected waves at ω_S can only be studied well if the medium is absorbing at ω_S. If this is the case, there may still be a transmitted wave at ω_S emerging from the slab, when the fundamental or composing frequencies are all transmitted. In this situation $\mathbf{k}_T$ is complex, but $\mathbf{k}_S$ is not. There is a uniform polarization source wave set up in the medium. From Eq. (4-3) one finds immediately a transmitted intensity, when the homogeneous wave has decayed with complex ϵ_T,

$$|E_\perp^T|^2 = 16\pi^2 |P_\perp^{NLS}|^2 |\epsilon_T - \epsilon_S|^{-2} \tag{4-15}$$

Appendix 2 gives a complete discussion for the case of the other polarization. It also analyzes the phenomena of total reflection and total transmission in the nonlinear case. Even if ϵ_R and ϵ_T are real, the angles of incidence may be chosen such that one or more of the vectors $\mathbf{k}_R$, $\mathbf{k}_S$, and $\mathbf{k}_T$ become complex. A special case of second harmonic reflection with polarization in the plane of reflection is presented by the nonlinearity in a plasma of free electrons, discussed in section 1-2. Kronig and Boukema[4] have treated this case.

Anisotropic Media

Similar, but algebraically more complex, arguments can be given for uniaxial and biaxial optical crystals. These crystals are of great interest because they sometimes permit the matching of phase velocities between the fundamental and harmonic waves. Kleinman has discussed this case. Incident waves will in general be refracted into two waves. These will in general not obey Snell's law, but the tangential components of the wave vectors along the boundary are continuous. The refracted waves create nonlinear source waves. Take one with wave vector $\mathbf{k}_S$ and polarization direction $\hat{p}$, and magnitude $|P^{NLS}|$. There are in general two wave vectors $\mathbf{k}_\nu$ with $\nu = 1,2$ with the same tangential components as $\mathbf{k}_S$. Assume that these components vanish. Then $\mathbf{k}_1$, $\mathbf{k}_2$, and $\mathbf{k}_S$ are all in the normal direction $\hat{n}_S$. The polarization vectors of the waves are $\hat{e}_1$ and $\hat{e}_2$. The homogeneous wave equation has the form

$$(c/\omega)^2 |k_\nu|^2 [\hat{e}_\nu - \hat{n}_S(\hat{e}_\nu \cdot \hat{n}_S)] - \boldsymbol{\epsilon} \cdot \hat{e}_\nu = 0 \tag{4-16}$$

for $\nu = 1,2$.

One can also define three unit vectors $\hat{d}_1$, $\hat{d}_2$, and $\hat{d}_3$ such that $\hat{d}_1$ is parallel to $\epsilon \cdot \hat{e}_1$, $\hat{d}_2$ is parallel to $\epsilon \cdot \hat{e}_2$, but $\hat{d}_3$ is taken orthogonal to $\hat{d}_1$ and $\hat{d}_2$. ($\hat{d}_3 = \hat{e}_3 = \hat{n}^S$). It is well known that $\hat{d}_1 \cdot \epsilon \cdot \hat{e}_2 = \hat{d}_2 \cdot \epsilon \cdot \hat{e}_1 = \hat{d}_3 \cdot \epsilon \cdot \hat{e}_1 = \hat{d}_3 \cdot \epsilon \cdot \hat{e}_2 = 0$. With

$$\mathbf{A} = \hat{e}_1 A_1 + \hat{e}_2 A_2 + \hat{e}_3 A_3 \tag{4-17}$$

Eq. (4-4) becomes

$$\sum_{\nu=1}^{2} \left\{ \left(\frac{c}{\omega}\right)^2 |k^S|^2 [\hat{e}_\nu - n^S(\hat{n}^S \cdot \hat{e}_\nu)] - \epsilon \cdot \hat{e}_\nu \right\} A_\nu - \epsilon \cdot \hat{e}_3 A_3 = 4\pi \hat{p} P_0$$

and on substitution of Eq. (4-16)

$$\sum_{\nu=1}^{2} \left\{ \left(\frac{c}{\omega}\right)^2 [|k^S|^2 - |k_\nu|^2][\hat{e}_\nu - \hat{n}^S(\hat{n}^S \cdot \hat{e}_\nu)] \right\} \times A_\nu - \epsilon \cdot \hat{e}_3 A_3 = 4\pi \hat{p} P_0 \tag{4-18}$$

Since $\hat{d}_3 = \hat{n}^S = \hat{e}_3$, the inner product with $\hat{n}^S$ determines A_3;

$$A_3 = -4\pi(\hat{n}_3 \cdot \hat{p}) P_0 / (\hat{n}_3 \cdot \epsilon \cdot \hat{n}_3) \tag{4-19}$$

Eliminating A_3 from Eq. (4-19)

$$\sum_{\nu=1}^{2}\left\{\left(\frac{c}{\omega}\right)^{2}[\,|k^{S}|^{2} - |k_{\nu}|^{2}\,][\,\hat{e}_{\nu} - \hat{n}^{S}(\hat{n}^{S}\cdot\hat{e}_{\nu})]\right\}$$

$$\times A_{\nu} = 4\pi P_{0}\left[\hat{p} - \frac{\epsilon\cdot\hat{n}^{S}(\hat{n}^{S}\cdot\hat{p})}{(\hat{n}^{S}\cdot\epsilon\cdot\hat{n}^{S})}\right] \tag{4-20}$$

and on taking the inner product with $\hat{d}_1$ or $\hat{d}_2$ (see Eq. (4-16) for $\nu = 1$ or 2)

$$A_{\nu} = \frac{4\pi P_0(\omega^2/c^2)[(\hat{d}_{\nu}\cdot\hat{p}) - (\hat{d}_{\nu}\cdot\epsilon\cdot\hat{n}^{S})(\hat{n}^{S}\cdot\hat{p})/(\hat{n}^{S}\cdot\epsilon\cdot\hat{n}^{S})]}{[\,|k^{S}|^{2} - |k_{\nu}|^{2}\,](d_{\nu}\cdot\hat{e}_{\nu})} \tag{4-21}$$

The complete solution is made up of the inhomogeneous solution and the two homogeneous waves. From the properties of the linear medium their wave vectors and polarization vectors are known. Their amplitudes must be determined, together with the amplitude and direction of polarization of the normally reflected wave, from the continuity equations for E_x, E_y, H_x, H_y, at $z = 0$.

4-2 COUPLING BETWEEN TWO WAVES: HARMONIC GENERATION

It has been shown that the phase-matching condition can be satisfied for the fundamental wave and a harmonic in anisotropic crystals. The phase-matching condition can also be satisfied in isotropic materials when regions of anomalous dispersion are included in the frequency span of the waves. It can also be satisfied in isotropic media with normal dispersion if more than three waves are present. It can also be satisfied for combinations of optical and acoustical waves. For these reasons it is important to investigate the nature of the solution, when the denominator in Eq. (4-9) blows up and the intensity according to Eq. (4-12) blows up as z^2.

Consider the simplest but important case of second harmonic generation. Denote the frequencies by ω_1 and $\omega_2 = 2\omega_1$, the polarization vectors by $\hat{e}_1$ and $\hat{e}_2$, the wave vectors of the homogeneous waves by $\mathbf{k}_1$ and $\mathbf{k}_2$. Again assume a plane boundary of the nonlinear medium at $z = 0$. The tangential components satisfy $k_{2x} = 2k_{1x}$, $k_{2y} = 2k_{1y}$, but due to dispersion $k_{2z} - 2k_{1z} = \Delta k$. For generality we admit $\Delta k \neq 0$, because the interest will also be in the neighborhood of perfect matching. We are guided by the small signal solution in which we found $\mathbf{E}_2(2\omega) = \hat{e}_2 A_2(z)\exp(+i\mathbf{k}_2\cdot\mathbf{r})$. If $A_2(z)$ becomes comparable in magnitude to A_1, we expect on the basis of energy considerations that A_1 will decrease and cannot be considered any longer as a fixed parameter. Waves at all other harmonic frequencies will be ignored. They will have a much larger phase mismatch and according to Eq. (4-13) these waves will represent rapidly varying perturbations of very small amplitude. Thus, only the following two wave equations of the type (3-23) are considered,

$$\begin{aligned} \nabla \times \nabla \times \mathbf{E}_2 - |k_2|^2 \mathbf{E}_2 &= 4\pi(2\omega/c)^2 \mathbf{P}^{NLS}(2\omega) \\ \nabla \times \nabla \times \mathbf{E}_1 - |k_1|^2 \mathbf{E}_1 &= 4\pi(\omega/c)^2 \mathbf{P}^{NLS}(\omega) \end{aligned} \tag{4-22}$$

Try the solution of the form

$$\begin{aligned} \mathbf{E}_1(\omega) &= \hat{e}_1 A_1(z) \exp(+i\mathbf{k}_1 \cdot \mathbf{r}) \\ \mathbf{E}_2(2\omega) &= \hat{e}_2 A_2(z) \exp(+i\mathbf{k}_2 \cdot \mathbf{r}) \end{aligned} \tag{4-23}$$

Substitution gives two second-order differential equations in A_1 and A_2. If the variation of the amplitude over one wave length is small, $|\partial^2 A/\partial z^2| \ll k|\partial A/\partial z|$, the second-order derivatives may be dropped and two coupled amplitude equations result

$$\begin{aligned} &-2ik_1 \cos\alpha_1 \cos(\alpha_1 - \beta_1) \frac{\partial A_1}{\partial z} \\ &\quad = \frac{4\pi\omega^2}{c^2} \hat{e}_1 \cdot \boldsymbol{\chi}(\omega = 2\omega - \omega) : \hat{e}_2 \hat{e}_1 A_2 A_1^* e^{i\Delta kz} \end{aligned} \tag{4-24}$$

$$\begin{aligned} &-2ik_2 \cos\alpha_2 \cos(\alpha_2 - \beta_2) \frac{\partial A_2}{\partial z} \\ &\quad = \frac{4\pi(2\omega)^2}{c^2} \hat{e}_2 \cdot \chi(2\omega = \omega + \omega) : \hat{e}_1 \hat{e}_1 A_1^2 e^{-i\Delta kz} \end{aligned} \tag{4-25}$$

The angle α is the angle between **D** and **E** in the anisotropic medium. It is also the angle between the wave vector **k** and the Poynting vector. The angle β is the angle between the wave vector and the boundary normal $\hat{z}$, as shown in Figure 4-5. The nonlinear susceptibilities are derivable from the enthalpy,

$$F = 2\,\mathrm{Re}(\chi E_1^2 E_2^*) \tag{4-26}$$

and satisfy the symmetry relation

$$2\chi_{ijj}(2\omega = \omega + \omega) = \chi_{jij}(\omega = 2\omega - \omega).$$

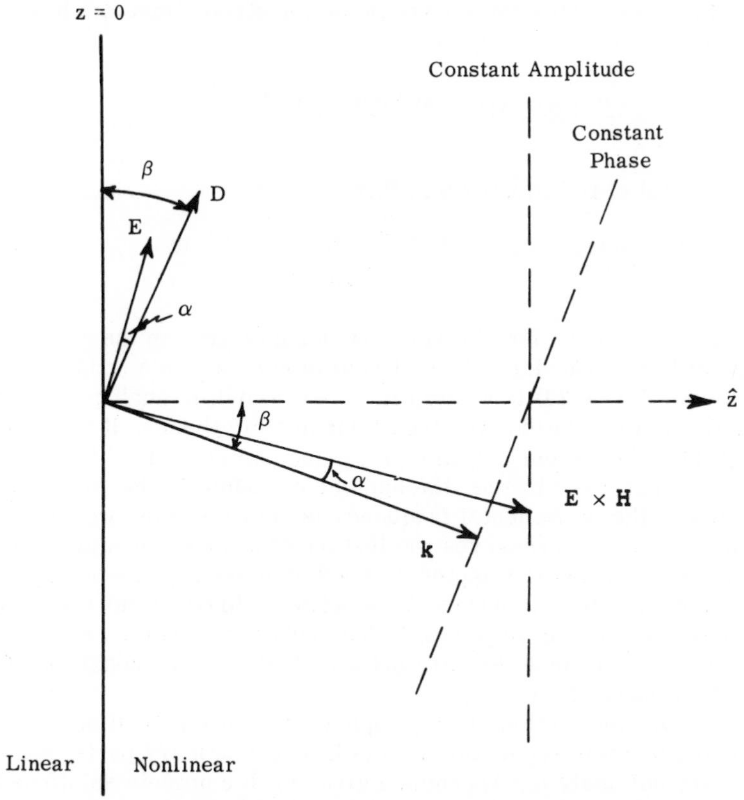

Figure 4-5. The parametric generation of a wave in an anisotropic medium, showing the angles between wave vector, Poynting vector, and boundary normal. This figure applies to a positive uniaxial crystal ($\alpha > 0$). In a negative uniaxial crystal, such as KDP, the vector **E** lies between **D** and **k** ($\alpha < 0$).

One integral of the two coupled nonlinear equations is readily obtained. Multiply Eq. (4-25) by $A_2^* c^2/8\pi\omega^2$, multiply the complex conjugate of Eq. (4-24) by $-A_1 c^2/4\pi\omega^2$, and add. Because of the symmetry relations the right-hand sides cancel.

$$k_1 \cos \alpha_1 \cos(\alpha_1 - \beta_1) A_1 \frac{\partial A_1^*}{\partial z} + \tfrac{1}{2} k_2 \cos \alpha_2 \cos(\alpha_2 - \beta_2) \times A_2^* \frac{\partial A_2}{\partial z} = 0 \tag{4-27}$$

The Poynting vector for a wave in an anisotropic medium has a magnitude

$$|S| = \frac{c}{4\pi}|\mathbf{E}\times\mathbf{H}| = \frac{kc^2\cos\alpha\, AA^*}{2\pi\omega} \tag{4-28}$$

An integral of the equations is thus

$$\frac{|S_1|\cos(\alpha_1-\beta_1)}{\omega} + \frac{2|S_2|\cos(\alpha_2-\beta_2)}{2\omega} = \frac{W}{\omega} \tag{4-29}$$

W is precisely the total power flow per unit area in the z-direction, normal to the boundary. It is of course constant in a nondissipative medium. Eq. (4-29) is written in a form which is analogous to the Manley-Rowe relations in circuit parametric theory. If the number of quanta passing per centimeter square per second in the z-direction at 2ω is increased by one, the number of quanta in the power flux density at the fundamental frequency is decreased by two. The classical theory thus preserves the feature of a three-quantum scattering process discussed in Chapter 2. Manley-Rowe integral relationships are also valid for the interactions between three or more waves. Examples are given in Appendix 1. The relations are then even more informative, because they are not identical with the condition of constant total power flow.

A complete solution of the coupled differential equations can only be given by their separation into real and imaginary parts, because they are not analytic. Appendix 1 gives such complete solutions for two coupled amplitude equations in section 5 and for three coupled waves in section 6. Here only one particular solution for exact phase matching between the fundamental and second harmonic will be presented.

Inspection of Eqs. (4-24) and (4-25) shows that for $\Delta k = 0$, one may choose $A_1 = \rho_1$ real, and $A_2 = i\rho_2$ pure imaginary. The real amplitudes satisfy the differential equations for $\alpha_1 = \alpha_2 = 0$, $\beta_1 = \beta_2 = 0$, $k_2 = 2k_1$,

$$\frac{d\rho_1}{dz} = -C\rho_1\rho_2 \qquad \frac{d\rho_2}{dz} = C\rho_1^2 \tag{4-30}$$

with

$$C = 2\pi\omega^2(\hat{e}_1\cdot\chi(\omega)\hat{e}_2\hat{e}_1)/k_1c^2 \qquad \alpha_1 = \alpha_2 = 0,\ \beta_1 = \beta_2 = 0 \tag{4-31}$$

Assume that initially $\rho_1 = \rho_1(0)$ and $\rho_2 = 0$ for $z = 0$. The actual boundary condition for the harmonic wave with amplitude $\rho_2(0)$ is of course slightly different as shown by Eq. (4-10). The initial amplitude of ρ_2 is so small that the build-up of the second harmonic is well represented by this condition. One has of course the power flow integral

$$\rho_1^2 + \rho_2^2 = \rho_1^2(0)$$

Substitution into the last of Eqs. (4-30) gives

$$\frac{d\rho_2}{dz} = C\{\rho_1^2(0) - \rho_2^2\} \tag{4-32}$$

An elementary integration gives the solution

$$\rho_2 = \rho_1(0) \operatorname{tgh} C\rho_1(0)z \tag{4-33}$$

As the second-harmonic intensity builds up, the power at the fundamental frequency is depleted as

$$\rho_1 = \rho_1(0) \operatorname{sech} C\rho_1(0)z \tag{4-34}$$

A characteristic interaction length may be defined by

$$\ell = C^{-1}\rho_1^{-1}(0) = \{2\pi\epsilon^{-1}k\chi(\omega)\rho_1(0)\}^{-1} = \frac{n\lambda}{8\pi^2\chi(2\omega)\rho_1(0)} \tag{4-35}$$

In this length about sixty per cent of the power would be converted to second harmonic; n is the index of refraction, λ is the vacuum wave length. When numerical values are substituted, it must be remembered that in our definition of the amplitudes $\chi(2\omega)$ is twice as large as the value usually quoted in the literature, but the amplitude $\rho_1(0)$ is twice as small for given power level. The coherence length is of course independent of the choice. With our definition $\chi(2\omega) = 6 \times 10^{-9}$ esu for KDP. For the amplitude $2\rho_1(0) = 2.7 \times 10^5$ volts/cm, corresponding to a power flux density $nc\,\rho_1^2/2\pi = 100$ megawatt/cm², one finds $\ell = 0.5$ cm for $\lambda = 6940$ Å.

Terhune and coworkers[5] have been able to convert more than 20 per cent of the light of a ruby laser beam to the second-harmonic frequency. The beam was slightly focused and carefully aligned along the phase matched direction in a KDP crystal. The fact that not all power could be converted is probably caused by the presence of several modes (waves) in the laser beam which are not simultaneously phase matched. It is a widespread misconception that for "reasons of a thermodynamic equilibrium" not more than about fifty per cent could be converted. If the phase of the incident wave is completely defined its entropy is essentially zero and in the coherent process without randomization this remains so. Thermodynamics is irrelevant.

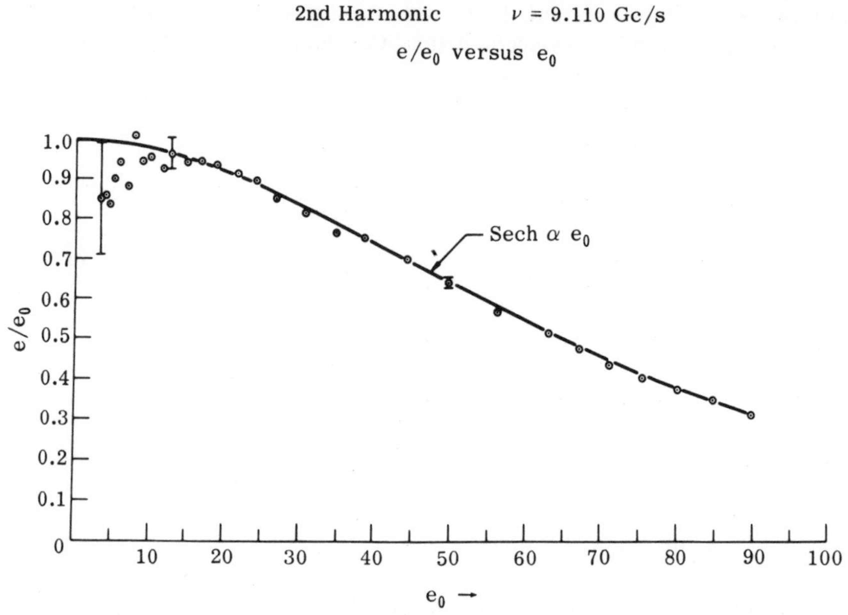

Figure 4-6. The attenuation of the acoustic amplitude at 9.1×10^9 cps, as a function of the initial amplitude e_0, due to second harmonic generation in a MgO crystal. The drawn curve is predicted by Eq. (4-34). (Courtesy of N. S. Shiren.)

The importance of Eq. (4-33) is that it predicts that complete conversion is possible. Although this has not been achieved for light harmonic generation, it has been obtained by Shiren[6] for acoustic waves. His experimental points are compared with the theoretical sech curve for depletion of the fundamental amplitude as a function of the initial amplitude in Figure 4-6. The next section will show that the same theory indeed applies to acoustic nonlinearities.

4-3 INTERACTIONS WITH VIBRATIONAL WAVES

Acoustic Nonlinearities

Anharmonic terms in the elastic properties of solids have been known for a long time. The theory of vibrational waves can be and has been extended into the nonlinear domain in a similar manner, as has been done for the electromagnetic waves. Since the acoustic vibrations admit three modes of polarization for a given direction of propagation, and the stress-strain relationship is described by a fourth-order tensor already in the linear domain, the theory of acoustic vibrations is more complex. Here the interest is not in a complete treatment of acoustical waves, but to emphasize the acoustic nonlinear properties and their similarity with the electromagnetic nonlinearity. For this purpose it will suffice to consider only pure longitudinal acoustical modes, propagating in the z-direction. At microwave frequencies the acoustical wave length is comparable to optical wave lengths. The acoustic dispersion is still very small, and perfect phase matching can be maintained over distances of 1 cm. Furthermore, the acoustical mean free path in pure crystals at liquid helium temperature can be 10 cm or more. The waves can readily be excited to intensities at which nonlinearities come into play. The relationship between P and E is now taken over by the stress σ and the strain S. The linear elastic moduli are denoted by C, the lowest order nonlinear ones by A. The general tensor relationship is,

$$\sigma_{ij} = C_{ik}^{j\ell} S_{k\ell} + A_{ikm}^{j\ell n} S_{k\ell} S_{mn} + \cdots \tag{4-36}$$

For our case the interest will be limited to

$$\sigma_{zz} = C S_{zz} + C^{NL} S_{zz}^2 \tag{4-37}$$

where C^{NL} has been written for A_{zzz}^{zzz} and $C = C_{zz}^{zz}$. Quite generally the acoustic nonlinearity[7] is such that $3C < C^{NL} < 16C$. The equation of motion for a longitudinal wave propagating in the z-direction is derived by considering the equation of motion of a unit volume with mass density ρ and deviation u in the z-direction. With the strain $S_{zz} = \partial u/\partial z$, the linear wave equation is,

$$\rho \ddot{u} = \frac{\partial \sigma_{zz}}{\partial z} = C \frac{\partial^2 u}{\partial z^2} \tag{4-38}$$

or for the dimensionless strain,

$$\ddot{S} - (C/\rho)\partial^2 S/\partial z^2 = 0 \tag{4-39}$$

The proper dispersion law $C(\omega)$ can be found in standard texts on lattice vibrations,[7,8] but $C(\omega)$ changes by less than $1:10^6$ between 0 and 10^{10} cps. Solutions of the linear wave equation can be written in the form, $2\,\mathrm{Re}\,\{S_1 \exp(iq_1 z - i\omega_1 t)\}$, etc., with a phase velocity $\omega_1/q_1 = (C/\rho)^{1/2}$. The nonlinear wave equation is

$$\ddot{S} - (C/\rho)\,\partial^2 S/\partial z^2 = (C^{NL}/\rho)\,\frac{\partial^2 (S^2)}{\partial z^2} \tag{4-40}$$

One recognizes an inhomogeneous nonlinear source term which for strains, $S < 10^{-2}$, may be considered as a small perturbation. If an acoustic wave S_1 at ω_1 is present, a second-harmonic wave will be generated through the term in S_1^2. The growth of $S_2(z)$ may be described by the coupled amplitude equations, provided $|\partial^2 S/\partial z^2| \ll q|\partial S/\partial z| \ll q^2|S|$,

$$-(C/\rho)(2iq_1)\frac{\partial S_1}{\partial z} = (C^{NL}/\rho)(-2q_1^2)S_2 S_1^* \tag{4-41}$$

$$-(C/\rho)(2iq_2)\frac{\partial S_2}{\partial z} = (C^{NL}/\rho)(-4q_1^2)S_1^2 \tag{4-42}$$

These equations have exactly the same form as the Eqs. (4-24) and (4-25) for the electromagnetic case for perfect matching. There is, however, an important distinction. In the acoustic case the system is not limited to two waves, because in general the third harmonic is also phase matched. This increases the complexity of the problem enormously. One must not only add an equation for the third harmonic

$$-(C/\rho)(2iq_3)\partial S_3/\partial z = (C^{NL}/\rho)(-9q_1^2)S_1 S_2 \tag{4-43}$$

but also add a term proportional to $S_3 S_2^*$ to Eq. (4-41) and a term $S_3 S_1^*$ to Eq. (4-42). As soon as the third harmonic has reached an appreciable intensity, the fourth harmonic will be created, etc. It is fortunately possible to change the propagation characteristics of an ultrasonic wave by its interaction with paramagnetic ions. Their spin levels can be tuned to an acoustic frequency by an external magnetic field. This causes absorption and anomalous dispersion for the ultrasonic wave. Such experiments have been carried out by Shiren in an MgO crystal containing Ni^{++} or Fe^{++} ions. Shiren was able to suppress the third harmonic in this manner. Mathematically the effect of the magnetic ions may be described by adding to the elastic modulus a complex quantity ΔC_{magn} in the wave equation. In the amplitude equation the magnetic perturbation is described by the addition of a term $\Delta C_{magn}\rho^{-1}q_3^2 S_3$ to the right-hand side of Eq. (4-43). The real part of ΔC_{magn} causes a phase mismatch, the imaginary part an attenuation of the wave S_3. Near magnetic resonance both causes will contribute to keep the amplitude of S_3 down. Consequently the Eqs. (4-41) and (4-42) are a good approximation to the physical situation in this case. In this manner Shiren[6] obtained the experimental points in Figure 4-6, which confirm the theoretical sech behavior for the depletion of the fundamental amplitude in a striking manner. When the third harmonic is not suppressed, the fundamental decays more slowly as a function of initial amplitude. In this nondispersive case all harmonics are generated and the solution should be described by equations of the same type as Eqs. (3-27) and (3-29).

Brillouin Scattering

There is also a coupling between the acoustic waves and light waves. If a material is elastically deformed, the index of refraction changes. This effect is well known and used in optical stress analysis. The change of index of refraction with density under hydrostatic compression has also been known for a long time. Brillouin systematically investigated the behavior of light waves in the presence of periodic elastic deformations. The effect may be described as a diffraction of a light wave by a variable index of refraction grating, set up by the acoustic vibration, or an incident light wave with wave vector $\mathbf{k}_1$ and frequency ω_1 collides with an acoustic wave with the wave vector $\mathbf{q}_2$ and frequency ω_2 to give a light wave with wave vector $\mathbf{k}_3$ and frequency ω_3. The conservation of energy and quasi momentum requires $\omega_3 = \omega_1 + \omega_2$ and $\mathbf{k}_3 = \mathbf{k}_1 + \mathbf{q}_2$. For acoustical phonons with frequency below 10^{10} cps, these conditions can always be satisfied. Although the momentum q_2 can be comparable in magnitude to k_1 and k_2, the acoustical frequency is very small compared to the light frequency. The momentum triangles shown in Figure 4-7 may be considered to be isosceles, with $|k_1| \simeq |k_3|$. The Brillouin relation for the angle of scattering of the light follows immediately,

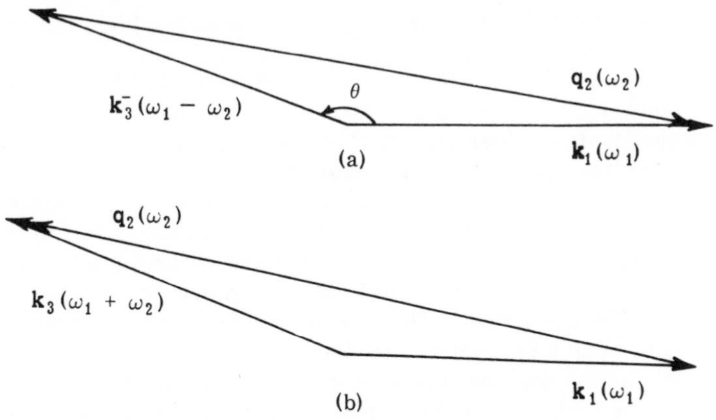

Figure 4-7. Brillouin scattering. In (a) a vibrational quantum is emitted, and a light quantum at a smaller frequency than the incident wave ω_1 is scattered. In (b) sound energy is absorbed. The scattered light wave has a larger frequency than the incident light wave at ω_1.

$$2k \sin\frac{\theta}{2} = q$$

or

$$\omega_{ac} = 2\omega_1(v/c)\sin\frac{\theta}{2} \tag{4-44}$$

where v/c is the ratio of the phase velocities of sound and light in the medium. The maximum acoustic frequency that can be involved in Brillouin scattering is obtained for backward scattered light, $\theta = \pi$. The intensity of the parametrically generated scattered light may be calculated in the same manner as the generation of the sum frequency by two other light waves.

The time-averaged energy of the medium in the presence of the two light waves and one acoustic wave is represented by

$$\langle F \rangle = 2 \operatorname{Re} p E_1 S_2 E_3^* \tag{4-45}$$

If the frequency $\omega_2 = 0$ and E_3^* is replaced by E_1^*, the term describes the photoelastic coupling energy. The quantity p is the photoelastic constant. It is really a fourth-rank tensor, connecting the strain tensor with two electric field vectors. The photoelectric constants have been measured in a number of materials and are typically of the order of unity, say $p = 0.5$.

From the time-average energy one immediately derives the polarization at ω_3,

$$P(\omega_3) = pE_1 S_2 \exp\{i(\mathbf{k}_1 + \mathbf{q}_2)\cdot\mathbf{r} - i(\omega_1 + \omega_2)t\} \tag{4-46}$$

The generation of the light wave at ω_3 is described by Eq. (4-12), which holds for the case of perfect momentum matching as long as the intensity at ω_3 does not build up to such a power level that depletion of either or both of the incident waves results.

An interesting recent application of the Brillouin scattering is its incorporation in a gas laser resonator system, as shown in Figure 4-8. The mirrors are totally reflecting and if the quartz crystal is not driven piezoelectrically, no light comes out. If a traveling ultrasonic wave is generated in the quartz crystal in the transverse direction and of such a frequency that the matching conditions are satisfied, light at the displaced frequency will be coupled out. The magnitude of the coupling can be adjusted by variation of the amplitude of the ultrasonic wave.

A light wave at the difference frequency, $\omega_3^- = \omega_1 - \omega_2$ can also be generated by a photoelastic polarization

$$P(\omega_3^-) = pE_1 S_2^* \exp\{i(\mathbf{k}_1 - \mathbf{q}_2)\cdot\mathbf{r} - i(\omega_1 - \omega_2)t\} \tag{4-47}$$

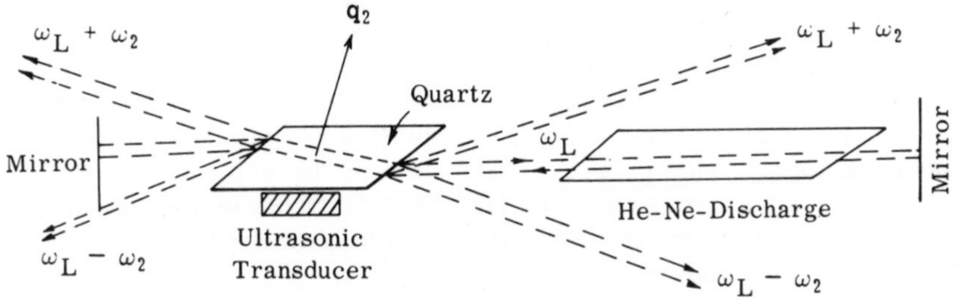

Figure 4-8. Ultrasonic coupling out of a gas laser with totally reflecting mirrors. Note the Brewster angles to avoid reflections. The ultrasonic traveling wave with wave vector $\mathbf{q}_2$ is absorbed after traversing the quartz crystal. (After A. E. Siegman, C. F. Quate, J. Bjorkholm, and G. Francois, App. Phys. Letters, **5**, 1 [1964].)

In this case the momentum matching requires

$$\mathbf{k}_3^- = \mathbf{k}_1 - \mathbf{q}_2$$

and the ultrasonic wave will grow together with the scattered light wave at ω_3^-, while the incident light wave will eventually be depleted.

If the light wave and/or the acoustic wave have a standing wave pattern, all possible combinations of forward and backward waves should be considered.

The same time-averaged photoelastic energy describes also the parametric generation of an acoustic wave as the difference beat between two light waves. Differentiation with respect to the strain amplitude gives a stress Fourier component at $\omega_2 = \omega_1 - \omega_3$

$$\sigma_{\text{photo}}(\omega_2) = pE_1E_{3-}^* \exp\{i(\mathbf{k}_1 - \mathbf{k}_3)\cdot\mathbf{r} - i\omega_2 t\} \tag{4-48}$$

For $\omega_1 = \omega_3$ one obtains a dc stress which is equivalent to the well-known electrostrictive effect for $\omega_1 = \omega_3 = 0$. The periodic stress at ω_2 serves as a source for the acoustic wave according to the inhomogeneous wave equation (compare Eq. (4-40),

$$\rho\ddot{S} - C\frac{d^2S}{dz^2} = \frac{d\,\sigma_{\text{photo}}}{dz^2}$$

$$= -|\mathbf{k}_1 - \mathbf{k}_3|^2 pE_1E_{3-}^* \exp\{i(\mathbf{k}_1 - \mathbf{k}_3)\cdot\mathbf{r} - i\omega_2 t\} \tag{4-49}$$

It has already been observed that the light wave at ω_{3^-} will grow at the same time as the acoustic wave at ω_2. This raises the interesting question what happens if the initial amplitude E_{3^-} is chosen smaller all the time. Both waves at ω_2 and ω_{3^-} will continue to grow at the expense of the wave at the largest frequency ω_1. Finally E_{3^-} is taken to be zero. Can the two waves of the smaller frequencies build up from the noise? This problem of parametric down conversion and oscillation will be treated in the next section. It is, of course, not limited to acoustic interaction and occurs also in the interaction between light waves alone.

4-4 PARAMETRIC DOWN CONVERSION AND OSCILLATION

Consider the second-harmonic generation of light with perfect phase matching described by the Eqs. (4-33) and (4-34). Assume that the phase of the second-harmonic wave at one point is retarded by 180° with respect to the fundamental wave by a slab of dispersive material. After that the waves are again in the phase-matched condition. But now inspection of Eq. (4-30) shows that the amplitude derivatives have changed sign. The second harmonic will now be depleted and the fundamental wave is regenerated.

The question of subharmonic generation is what happens when the initial condition is $A_1(\omega) = 0$, $A_2(2\omega) = A_2(0)$. The answer cannot immediately be given because of the asymptotic nature of the curves. Both derivatives appear to vanish, $dA_1/dz = dA_2/dz = 0$, but the process may build up because $A_1(\omega) = 0 + A_{Noise}(\omega)$. It will be shown that the material exhibits a positive gain at ω. Consequently noise will be amplified and with appropriate feedback a subharmonic oscillator will result. This was pointed out by Kingston.[9] Kroll[10] analyzed in more detail the down conversion of a light wave at ω_1 into two other light waves at ω_2 and $\omega_{3^-} = \omega_1 - \omega_2$.

The light field at frequency ω_1, which may be identified with the pump or laser frequency ω_L, will be regarded as a fixed parameter. The two coupled wave equations at the signal frequency $\omega_2 = \omega_S$ and idler frequency $\omega_3 = \omega_i$ are,

$$\nabla \times \nabla \times \mathbf{E}_S - \frac{\epsilon(\mathbf{k}_S)\omega_S^2}{c^2}\mathbf{E}_S = \frac{4\pi\omega_S^2}{c^2}\chi(\omega_S = \omega_L - \omega_i)E_L E_i^* \tag{4-50}$$

$$\nabla \times \nabla \times \mathbf{E}_i^* - \frac{\epsilon^*(\mathbf{k}_i)\omega_i^2}{c^2}\mathbf{E}_i^* = \frac{4\pi\omega_i^2}{c^2}\chi^*(-\omega_i = -\omega_L + \omega_S)E_L^* E_S \tag{4-51}$$

Find a solution in the form

$$E_s = A_s \exp(i\mathbf{k}_s \cdot \mathbf{r} - i\omega_s t)$$
$$E_i^* = A_i^* \exp\{i(\mathbf{k}_s - \mathbf{k}_L)\cdot \mathbf{r} + i\omega_i t\} \qquad (4\text{-}52)$$

Substitution of these expressions into the two wave equations leads to two homogeneous algebraic relations in E_s and E_i^*. Nonvanishing solutions only occur if the determinant is put equal to zero. The presence of a plane boundary at $z = 0$ makes it convenient to label solutions for each pair of values k_{xs} and k_{ys}. These values are real. The only complex component is k_z. Attenuation and gain only occur in the direction normal to the boundary. The determinantal equation takes the form,

$$(k_{zs}^2 + (k_{xs}^2 + k_{ys}^2) - (k_s^0)^2)\{(k_{zs} - k_{zL})^2 + (k_{xs} - k_{xL})^2 + (k_{ys} - k_{yL})^2 - (k_i^0)^2\} = \left(\frac{4\pi\omega_i\omega_s}{c^2}\right)^2 \chi^2 |E_L|^2 \qquad (4\text{-}53)$$

This is a fourth-power polynomial in k_{sz} for each given value of the tangential component $(k_{xs}^2 + k_{ys}^2)^{1/2}$. The roots of the equation with a negative imaginary part represent a coupled wave mode with positive gain. The quantities k_s^0 and k_i^0 are the wave numbers of the homogeneous solutions for free waves in the absence of E_L. If the medium is lossy at ω_s or ω_i, these wave numbers are complex and $(k_s^0)^2$ should be replaced by $(k_s^0)^2(1 + i\epsilon_s''/\epsilon_s')$ for the wave E_s. $(k_i^0)^2$ should be replaced by $(k_i^0)^2(1 - i\epsilon_i''/\epsilon_i')$ for the wave E_i^*.

In order to make progress with the discussion introduce the notation for the z-components of the free waves with prescribed tangential component $k_{zs}^m = \{(k_s^0)^2 - (k_{xs}^2 + k_{ys}^2)\}^{1/2}$, $k_{zi}^m = \{(k_i^0)^2 - (k_{xi}^2 + k_{yi}^2)\}^{1/2}$. Introduce the wave vector mismatch in the z-direction, $\Delta k = k_{zs}^m - k_{zL} + k_{zi}^m$. Consider only those values of the tangential components for which $\Delta k \ll k_{sz}^0$. The root of the equation also differs by a small amount $\Delta\kappa$ from k_{sz}^m. The quartic equation is then written in the form for the lossless case,

$$(+2(k_{sz}^m)\Delta\kappa + (\Delta\kappa)^2)\{-(2k_{zi}^m(\Delta k + \Delta\kappa) + (\Delta k + \Delta\kappa)^2\} = \left(\frac{4\pi\omega_i\omega_s}{c^2}\right)^2 |\chi|^2 |E_L|^2 \qquad (4\text{-}54)$$

There are two roots $\Delta\kappa$ which are small, and two roots $\Delta\kappa$ near $-2k^0_{sz}$. The latter two solutions belong to backward traveling waves opposite to the direction of the pump. The first two roots are obtained to a very good approximation by dropping the terms in $(\Delta\kappa)^2$ and $(\Delta k + \Delta\kappa)^2$ in each bracket. The equation becomes then a quadratic with the two roots,

$$\Delta\kappa = -\tfrac{1}{2}\Delta k \pm \left[\tfrac{1}{4}(\Delta k)^2 - \left(\frac{4\pi\omega_i\omega_s}{c^2}\right)^2 \frac{1}{4k^m_{zs}k^m_{zi}} |\chi|^2 |E_L|^2\right]^{1/2} \tag{4-55}$$

It is seen that one root will have a negative imaginary part if

$$|\Delta k| < \frac{4\pi\omega_i\omega_s}{c^2(k^m_{zs}k^m_{zi})^{1/2}} |\chi E_L|$$

The gain is maximum for $\Delta k = 0$. The wave that has gain consists of a linear combination of the signal and idler waves. The ratio of the coefficients can be found as the eigenvector belonging to the eigenvalue $\Delta\kappa$. If the gain is larger than unavoidable losses due to scattering or reflection losses from mirrors that are used to provide feedback, a parametric oscillator will result. The requirement on the phase match is quite stringent, and it is important that the pump contains essentially only one direction, i.e., is diffraction limited.

The treatment may be extended to include damping in the manner indicated above. It can also be shown that the reduction of the quartic equation to a quadratic is equivalent to the coupled amplitude approximation of section 4-2. The damping can also be incorporated in this approximation.

In analogy to Eqs. (4-24) and (4-25) we have, dropping the angles, $\alpha_1 = \alpha_2 = 0$,

$$\frac{dE_s}{dz} = i\frac{4\pi\omega_s^2}{2k^m_{zs}c^2}\chi E_L E_i^* e^{i\Delta kz} - \alpha_s E_s \tag{4-56}$$

$$\frac{dE_i^*}{dz} = -i\frac{4\pi\omega_0^2}{2k^m_{zi}c^2}\chi^* E_L^* E_s e^{-i\Delta kz} - \alpha_i E_i^* \tag{4-57}$$

The amplitude attenuation coefficients α_s and α_i have been introduced. They are directly related to the imaginary part of the dielectric constant. The solution of these equations gives two waves with wave vector $k^m_{sz} + \Delta\kappa$,

$$\Delta\kappa = -\tfrac{1}{2}\Delta k + \tfrac{1}{2}i(\alpha_s + \alpha_i) \pm \left[\tfrac{1}{4}\{\Delta k - i(\alpha_i - \alpha_s)\}^2 - \left(\frac{4\pi\omega_i\omega_s}{c^2}\right)^2 \frac{|\chi|^2|E_L|^2}{4k_{zi}^m k_{zs}^m}\right]^{1/2} \tag{4-58}$$

In the absence of loss this solution is identical with Eq. (4-55). For perfect phase matching and equal loss for the two waves the parametric gain factor must be able to overcome the loss to obtain amplication

$$\frac{2\pi\omega_i\omega_s}{c^2(k_{zi}^m k_{zs}^m)^{1/2}}\chi|E_L| > \alpha_s = \alpha_i$$

This was of course to be expected. A very interesting result is obtained, however, when the damping of one wave is very much larger than that in the other wave. Take

$$\Delta k = 0, \quad \alpha_i \gg \frac{2\pi\omega_i\omega_s}{c^2(k_{zi}^m k_{zs}^m)^{1/2}}\chi|E_L|$$

In that case the square root may be expanded to give the roots

$$\Delta\kappa = +i\alpha_i + i\left(\frac{4\pi\,\omega_i\omega_s}{c^2}\right)^2 \frac{\chi^2|E_L|^2}{4k_{zi}^m k_{zs}^m\,\alpha_i} \tag{4-59}$$

$$\Delta\kappa = i\alpha_s - i\left(\frac{4\pi\omega_i\omega_s}{c^2}\right)^2 \frac{\chi^2|E_L|^2}{4k_{zi}^m k_{zs}^m\,\alpha_i} \tag{4-60}$$

If α_s is sufficiently small, there will always be gain in one wave, even though the other wave is very strongly attenuated. The amplified wave has almost pure ω_s character. A wave at ω_i with much smaller amplitude is, however, dragged along. This solution is most easily obtained by omitting the term $|dE_i^*/dz|$ compared to $\alpha_i E_i^*$ in Eq. (4-57). Substitution of E_i^* into Eq. (4-56) gives immediately the amplified wave corresponding to Eq. (4-60). When the solution of E_s is inserted back into Eq. (4-57), the polarization wave at ω_i is obtained. The ratio of the amplitudes is

$$|E_i/E_s| \approx \frac{4\pi\omega_i^2\chi|E_L|}{2k_i c^2\alpha_i}$$

These limiting cases are instructive. Quite generally, the imaginary part of $\Delta\kappa$ in Eq. (4-58), or even better, of the root $\Delta\kappa$ solved from a quartic equation such as Eq. (4-53) with damping added, gives a complete description of the gain or loss in all situations of parametric down conversion. For each eigenvalue $\Delta\kappa$ one may find the appropriate eigenwaves, i.e., the proper linear combination of the waves at ω_s and ω_i.

The same considerations are valid if one of the waves, say ω_i, is an acoustic wave. This is the case of Brillouin scattering. Eq. (4-50) is then replaced by the acoustic wave equation (4-49) which has an identical form. With the proper substitution of the physical quantities all formulae remain valid in this case. In crystals and fluids at room temperature the acoustic damping is much larger than the damping of the light wave in the transparent medium. A typical attenuation coefficient for an ultrasonic wave at 300°K is $\alpha_i = 400\ \text{cm}^{-1}$ at 10^{10} cps. α increases as the square of the frequency. The light attenuation is $\alpha_s < 0.1\ \text{cm}^{-1}$. One may thus expect gain to occur at the Brillouin scattered light frequency ω_s. The effect is very similar to the Raman effect. A laser quantum ω_L is absorbed, a quantum at $\omega_s = \omega_L - \omega_{ac}$ is emitted, while the acoustic quantum $\hbar\omega_{ac} = \hbar\omega_i$ is absorbed by the strong damping mechanism. It is thus seen that if α_i is gradually reduced from a large initial value to a small value comparable to α_s, the nature of the effect appears to change from a Raman-type process where the amplified wave has predominantly Brillouin-shifted light character, to the simultaneous parametric generation of a light wave and an acoustic wave. Experimentally this could possibly be accomplished by cooling the crystal to liquid helium temperature, where α_i is comparable to α_s.

It was shown in section 4-3 that the Brillouin scattered light may come out in the backward direction. This implies that k_{sz}^m and k_{sz} in Eqs. (4-58), (4-59), and (4-60) are negative. For a backward wave the attenuation coefficient α_s should be given the opposite sign. In the limit of large damping of the acoustic wave, the relation retains its validity for negative k_{zi} and negative α_s. The imaginary part of Δk changes its sign. Since the wave travels also in the opposite direction, the gain of the backward wave is still given by Eq. (4-60) if the vibrational damping is large. Kroll[11] has given a more general discussion of the stimulated Brillouin effect. He derives a formula for the gain and instability of the backward wave amplifier if the acoustic damping is not large compared to the parametric gain. He also shows that the assumption of a steady state solution may not be valid. During the duration of the laser pulse, the sound wave may not have had time to traverse the region of the crystal, over which significant gain occurs. Kroll has given

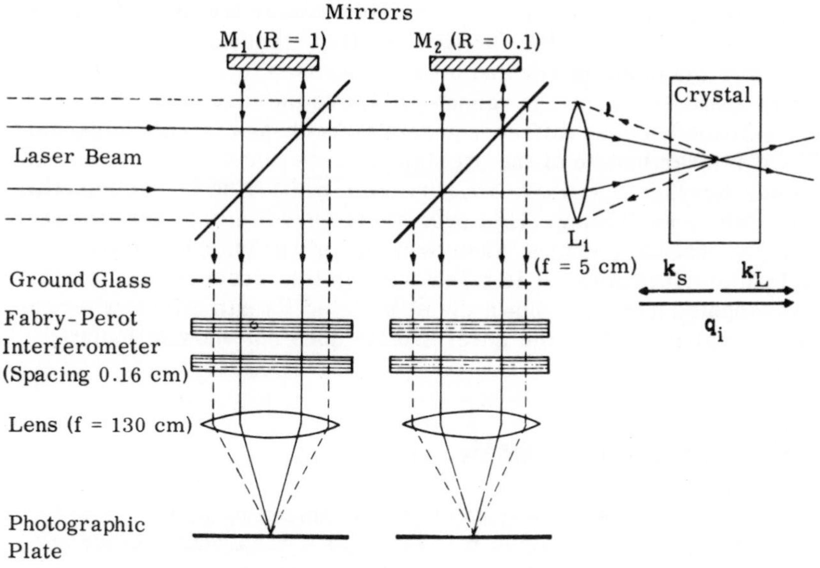

Figure 4-9. Experimental arrangement to detect the stimulated Brillouin scattering in the backward direction (after Chiao, Townes, and Stoicheff).[12]

a transient solution for this situation. Another theoretical discussion of the stimulated Brillouin effect has been given by Townes and coworkers.[12]

The effect has been observed by Chiao, Townes, and Stoicheff[12] in quartz and sapphire. The experimental arrangement is shown in Figure 4-9. The threshold at room temperature is quite high and requires an intense, strongly focused laser beam with estimated power flux densities in excess of 10^{10} watts/cm^2. A fraction ω_{ac}/ω_L of the laser power is converted into ultrasonic power and in less than 10^{-7} sec converted into heat by the damping mechanism. The crystals were invariably damaged when the intensity was high enough to observe the effect. The backward scattered light was observed. It probably has the lower threshold, because the path length over which gain can be obtained is clearly longer along the direction of the laser beam. It may be noted in this connection that forward Brillouin scattering is possible in anisotropic crystals. A light wave of one polarization may be scattered into a wave of the other polarization in the same direction with a lower frequency. Due to the birefringence it is possible to satisfy the conservation of energy and momentum.

Immediately above threshold, an appreciable fraction of the laser power is converted to the Brillouin shifted frequency. This depletes or limits the pump power and other processes, e.g., Brillouin scattering in other directions cannot get above the threshold. This competition between different possible processes is very important. The nonlinear nature of the problem is such that only a few with the lowest thresholds will go. The Brillouin scattering is in competition with Raman scattering. This process is very similar, with the role of the acoustical phonons taken over by optical phonons, e.g., molecular vibrations. It is noteworthy in this respect that quartz and sapphire have not shown the stimulated Raman effect, whereas calcite which exhibits the latter effects does not show stimulated Brillouin scattering.

4-5   STIMULATED   RAMAN   EFFECT

The theory of this effect has been discussed by many authors.[13-21] In a classical description of this effect as a parametric process[22] quite analogous to the Brillouin scattering, the coupling between a light wave at the stokes frequency ω_s and an optical phonon or vibrational wave at ω_v is produced by a pump field at $\omega_L = \omega_s + \omega_v$. The main distinction is that the dispersion law for optical phonons is very different from that of acoustical phonons. For the vibrational mode of a typical molecular group, say CO_3 in calcite or $C-H$ in a molecular organic fluid, the width of the corresponding phonon branch is very narrow. Since the interest is only in long wave phonons with wave length corresponding to the wave length of light waves, $ka < 10^{-3}$, the frequency ω_v is essentially constant. Therefore, a matching momentum $\mathbf{k}_v$ of the optical phonon can be found regardless of the direction in which the wave at ω_s travels. The dispersive behavior of the electromagnetic waves and the optical phonons is shown in Figure 4-10. Due to the vibrational-electronic coupling the curves do not cross. In fact there is an extremely narrow gap corresponding to the very strong absorption of an EM wave at the vibrational frequency. The Kramers-Kronig relations indicate that the dielectric constant for ω just above ω_v is slightly smaller than just below ω_v. The slope of the curves is proportional to $\epsilon^{-1/2}$. Further increase in frequency towards the visible part of the spectrum is accompanied by a decrease in slope corresponding to the natural color dispersion. To match the momenta for stokes light in the forward direction, one must have

$$k_L - k_s = \epsilon^{1/2}(\omega_L)\, c^{-1} (\omega_L - \omega_s) = \epsilon^{1/2}(\omega_L)\, c^{-1}\, \omega_v = k_v^{min}$$

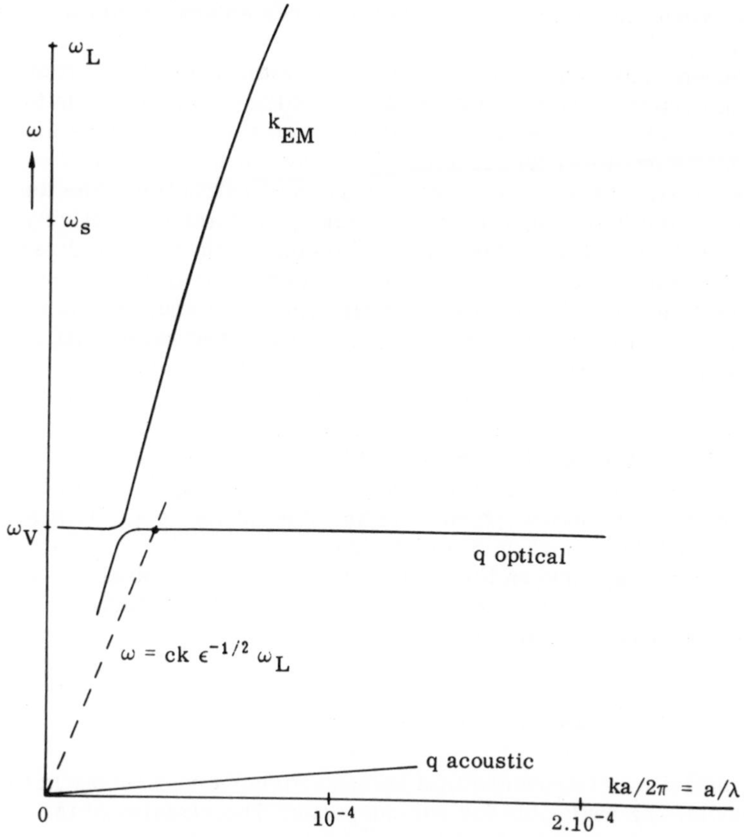

Figure 4-10. The dispersion of the electromagnetic waves, an optical phonon branch, and an acoustic phonon branch, for long wave lengths, $\lambda \gg a$. (The figure is not to scale.) Raman scattering from the sharp vibrational mode, with a relatively weak coupling to the electromagnetic field, is possible in the forward direction.

Such a phonon wave vector can be found, if $\epsilon^{1/2}(\omega_L) > \epsilon^{1/2}(\omega = \omega_V - \delta)$. This is possible as indicated by the dotted line with slope $\epsilon^{-1/2}(\omega_L)$. For stokes radiation in all other directions optical phonons with matching momenta larger than k_V^{min} can obviously be found. The dispersion of acoustical phonons is also indicated, but not to scale. Note that the whole picture is concerned only with very small values of ka, since the optical wave length is large compared to the interatomic spacing.

Loudon[16] has discussed the situation for the very strong photon-phonon coupling which occurs with ionic lattice vibrations. In this case the dielectric constant $\epsilon(\omega = \omega_V - \dot{0}) = \epsilon_{dc}$, whereas above the ionic vibrational frequencies it is very much smaller and equal to the square of the infrared optical index of refraction. The dispersive effects are much more pronounced and are qualitatively shown in Figure 4-11. Momentum matching in the forward direction of stokes radiation is now not possible in cubic or isotropic materials. Loudon gives a detailed treatment of the coupling between electromagnetic radiation and ionic vibrations and shows that in anisotropic crystals forward scattering is still possible.

Our discussion will be concerned with the much narrower optical phonon branches associated with molecular vibrations. The Raman effect has so far only been observed for these branches, presumably because they have less damping. The observation of the Raman effect on optical phonons in an ionic lattice, e.g., NaCl may be very difficult because of a high threshold.

The coupling between the nuclear vibrational coordinate R and the field is classically described by a polarizability which is parametrically dependent on R

$$\alpha = \alpha_0 + \frac{\partial \alpha}{\partial R} R + \cdots \tag{4-61}$$

This model was introduced by Placzek. Higher order terms in the expansion are negligible for our purposes. The enthalpy of the molecule in an electric field becomes

$$F = -\tfrac{1}{2}\alpha E^2 - \tfrac{1}{2}\frac{\partial \alpha}{\partial R} RE^2 \tag{4-62}$$

Instead of the localized vibration R, an optical phonon wave $Q = 2\,\mathrm{Re}\, A_V \exp(i\mathbf{k}_V \cdot \mathbf{r} - i\omega_V t)$ is introduced where $Q = R(2\rho\omega_V^2)^{1/2}$ is the normal coordinate of the vibrational wave, normalized per unit volume. The mass density is represented by ρ and ω_V is the vibrational frequency. In this manner a time averaged enthalpy per unit volume due to the simultaneous presence of two light waves and one optical phonon wave can be defined by

$$\langle F_{\mathrm{Raman}} \rangle = N(\partial\alpha/\partial Q)(A_V E_S E_L^* + A_V^* E_S^* E_L) \tag{4-63}$$

This expression has of course the same form as Eq. (4-45) for the photoelastic coupling. The coupled amplitude equations, analogous to Eqs. (4-56) and (4-57), can be written down. The wave equation

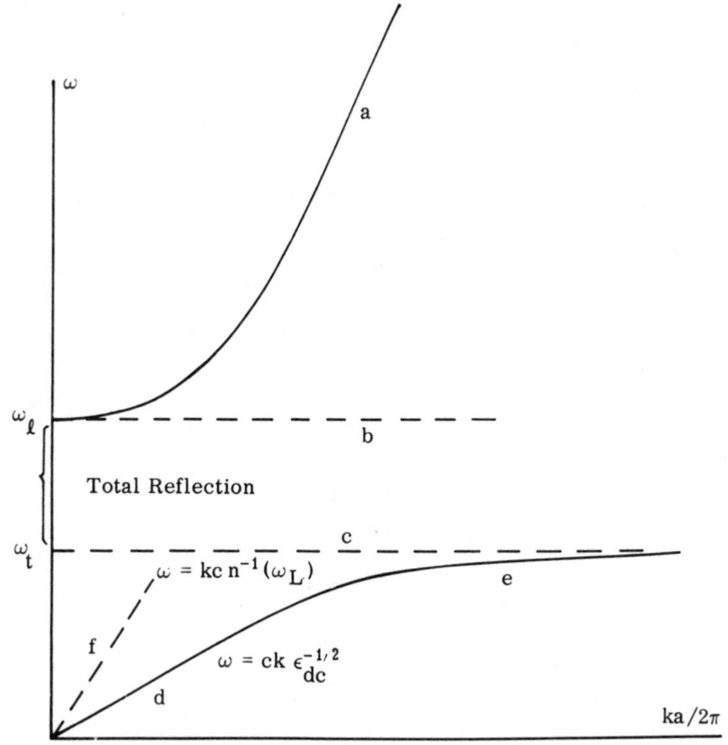

Figure 4-11. The dispersion of coupled electromagnetic waves and optically active ionic vibrations.
(a) Light wave.
(b) Longitudinal phonon } in absence of coupling.
(c) Transverse phonon } to EM waves.
(d) Low-frequency EM wave.
(e) Transverse ionic lattic phonon, coupled to d.
(f) Slope for matching of Raman scattering in forward direction. In this case of very strong coupling forward scattering is not possible in a cubic ionic crystal. Loudon[16] has shown that forward matching is possible in anisotropic crystals.

for optical phonons is different from Eq. (4-49) for acoustical phonons because of the different dispersion law. The group velocity $d\omega_v/dk_v$ therefore appears in the equation for the vibrational amplitude.

$$\frac{dE_s}{dz} = N(\partial\alpha/\partial Q)(i2\pi\omega_s^2/c^2 k_{sz}^m)E_L A_v^* \qquad (4\text{-}64)$$

$$\frac{dA_V^*}{dz} = N(\partial\alpha/\partial Q)\left\{-i\omega_V(dk_V/d\omega_V)\right\} E_L^* E_S - (dk_V/d\omega_V)\,\Gamma(k_V)A_V^* \qquad (4\text{-}65)$$

An essential feature is that the damping of the optical phonons is very much larger than that of the light waves. The damping in the latter is omitted. The attenuation coefficient in the optical phonon wave is written as $(dk_V/d\omega_V)\,\Gamma(k_V)$, where Γ has the dimension of a frequency and is the inverse of the characteristic damping time for the optical phonon. There should be no confusion with the α in Eq. (4-65) which stands for molecular polarizability, and the attenuation coefficients α_i and α_S in Eqs. (4-56) and (4-57).

There are again two exponential solutions of Eqs. (4-64) and (4-65) in which we have omitted the factor $\exp(i\Delta kz)$ because Δk can be taken as zero in any direction for $\omega_S = \omega_L - \omega_V$. They have a propagation constant in the z-direction $k_{Sz}^m + \Delta\kappa$ with

$$\Delta\kappa = +\tfrac{1}{2}i(dk_V/d\omega_V)\Gamma(k_V) \pm i\left[\tfrac{1}{4}(dk_V/d\omega_V)^2\Gamma_V^2 + \left\{(2\pi N^2\omega_S^2\omega_V(dk_V/d\omega_V)/c^2k_{Sz}^m\right\} \times (\partial\alpha/\partial Q)^2|E_L|^2\right]^{1/2} \qquad (4\text{-}66)$$

Because the damping Γ_V is always sufficiently large, the square root may be expanded, and a wave with gain is obtained

$$\Delta\kappa = -i(2\pi\omega_S^2/k_{Sz}^m c^2)\left\{N^2(\partial\alpha/\partial R)^2/2\rho\omega_V\Gamma(k_V)\right\}|E_L|^2 \qquad (4\text{-}67)$$

This wave corresponding to this eigenvalue is essentially a pure light wave at ω_S. A small admixture of the optical phonon wave is, however, still present. It represents the induced vibration by the simultaneous presence of the laser and stokes light waves. The amplitude of this vibrational wave follows immediately from Eq. (4-65) if dA_V^*/dz is ignored in comparison to the damping term.

It is of interest to compare this classical derivation of the stimulated Raman effect with the quantum mechanical calculation of the Raman susceptibility given by Eq.(2-62). In that case the vibrational levels were part of the quantized energy levels of a localized molecule. In the present treatment the quantized electric levels are hidden in the polarizability α, and the vibrations have been treated as an external classical wave. In terms of a Raman susceptibility the imaginary propagation constant representing gain is written as

$$\Delta\kappa = +i(2\pi\omega_S^2/k_{SZ}^m c^2)\,\chi_S''\,|E_L|^2 \tag{4-68}$$

or

$$\chi_S'' = -N^2(\partial\alpha/\partial R)^2/2\rho\omega_V\,\Gamma(k_V) \tag{4-69}$$

The equivalence of Eq. (2-62) with Eq. (4-69) will now be demonstrated. With the well-known expression for the linear susceptibility α_0 given, e.g., by the Kramers-Heisenberg expression Eq. (2-28), the imaginary part of Eq. (2-62) can be written in the form,†

$$\chi_S'' = -\frac{N}{\hbar\Gamma_{g'g^0}}\,\eta^2\alpha_0^2\,(\omega_V/\omega_L - \omega_{ng})^2 \tag{4-70}$$

The dimensionless constant $(\eta\omega_V/\omega_L - \omega_{ng})^2$ was the reduction factor for transitions connecting different vibrational levels. One therefore has the identification, $\eta\omega_V/\omega_L - \omega_{ng} = (R/\alpha_0)(\partial\alpha/\partial R)_0$. Furthermore, the classical energy of the vibrational oscillator with amplitude $2R$ is related to the frequency by

$$2m\omega_V^2 R^2 = 2\rho N^{-1}\omega_V^2 R^2 = \hbar\omega_V \tag{4-71}$$

Elimination of $\hbar$ and η from Eq. (4-70) gives complete agreement with Eq. (4-69).

The parallel between the quantum mechanical treatment and the classical treatment could be made even more complete, if the phonon normal modes were quantized rather than localized vibrational levels. An advantage of the nonlocalized treatment is that it shows explicitly that the damping constant may depend on the wave vector of the phonon. The damping is caused predominantly by the interaction with acoustical phonons via anharmonic elastic forces and crystalline imperfections. A detailed quantitative calculation of the decay of optical phonons into two or more acoustical phonons is difficult to carry out. On the basis of a general consideration of the available volume in momentum space for the final state, one may expect a phonon with a wave vector $k_L + k_S$ to be slightly more damped than a phonon with $k_L - k_S$. On this basis the gain per unit length for a stokes wave in the forward direction proportional to $\Gamma^{-1}(k_L - k_S)$ would be a few per cent higher than the gain per unit length of a stokes wave in the backward direction,

---

† It should be noted that the antiresonant term in Eq. (2-62) has the opposite sign from that occurring in the expression for the linear susceptibility α_0. Eq. (4-70) and expressions derived from it have therefore only an approximate character.

proportional to $\Gamma^{-1}(k_L + k_S)$. Bloembergen and Shen[22] have suggested this as a possible explanation for the observed forward-backward asymmetry in the intensity of the stokes radiation. With large exponential gain a difference of a few per cent in the damping could lead to an order of magnitude difference in intensity.

The process of elimination of the heavily damped phonon wave has led to a stokes gain independent of the phase of this wave. The growth of the stokes wave is described by the amplitude equation,

$$\frac{dE_S}{dz} = +(2\pi\omega_S^2/k_{Sz}c^2)\,|\chi_S''|\,E_S\,|E_L|^2 \tag{4-72}$$

The wave will continue to grow until the power level in the stokes light becomes an appreciable fraction of the incident laser power. Then E_L cannot any longer be considered as a fixed parameter. Its depletion should be considered according to the equation

$$\frac{dE_L}{dz} = -(2\pi\omega_L^2/k_{Lz}c^2)\,|\chi_S''|\,E_L\,|E_S|^2 \tag{4-73}$$

In contrast to the depletion of the pump power in parametric situations of harmonic generation, where it is an exception, depletion of the pump in Raman-type experiments is the rule and is unavoidable at high power levels or in long Raman cells. Loudon[16] has discussed the solution of the two coupled equations (4-72) and (4-73). Since the phases are not important, they can be transformed to a relation between intensities. Multiply Eq. (4-72) by E_S^* and Eq. (4-73) by E_L^*. Use the relationship between the number of photons passing through one cm^2 per second and $|E_S|^2$, $n_S\hbar\omega_S = (c\epsilon_S^{1/2}/2\pi)\,|E_S|^2$. There is of course a similar relationship at ω_i. The coupled amplitude equations may then be written in the form

$$\frac{dn_S}{dz} = W(n_S + 1)n_L, \quad \frac{dn_L}{dz} = -W(n_S + 1)\,n_L \tag{4-74}$$

with

$$W = \frac{8\pi^2\hbar\omega_S\omega_L}{c^2\epsilon_S^{1/2}\epsilon_L^{1/2}}\,|\chi''| \tag{4-75}$$

This form of the equations can be written down immediately by considering time proportional transition probabilities for Raman scattering. Hellwarth[13] and Zeiger[14] have used this form. At high power levels the +1 of spontaneous emission can be ignored. The equations may be integrated by elementary means. One integral is of course that the total number of photons is constant, $n_S + n_L = n_S^0 + n_L^0$.

The number of quanta at $z = 0$ are n_s^0 and n_L^0, respectively. The first equation (4-74) can be written as

$$\frac{dn_s}{dz} = Wn_s(n_s^0 + n_L^0 - n_s)$$

with the solution

$$n_s = \frac{n_s^0(n_s^0 + n_L^0)}{n_s^0 + n_L^0 \exp[-(n_s^0 + n_L^0)Wz]} \tag{4-76}$$

The range of validity of this solution is limited. First of all, there is not one stokes wave, but there is gain essentially in all directions. One particular direction can be singled out by mirrors lowering the threshold for that particular mode. This is done in a Raman laser. As soon as the stokes intensity has built up in this mode, it can in turn serve as a pump and create a wave at $\omega_{ss} = \omega_s - \omega_v = \omega_L - 2\omega_v$. These second-order stokes lines have been reported in the first publication on Raman lasers[23] and have been found by many other experimenters. In turn ω_{ss} can serve as a pump to create $\omega_{sss} = \omega_L - 3\omega_v$ and so on. The frequency ω_{ss} will not be created if the initial number of laser quanta is less than twice the threshold number. In that case n_s can at most build up to the threshold level for the second-order stokes process. The existence of a threshold due to damping could readily be incorporated in the rate Eq. (4-74).

Since the initial laser power can easily exceed the threshold by one or two orders of magnitude, a large set of coupled equations would have to be solved. The laser intensity drops rapidly to a value not more than a few times threshold, the stokes intensity becomes also limited by the second stokes process. What happens if the laser intensity is increased in a Raman laser is that more higher order stokes lines appear but that the intensity of each lower order stokes line cannot be pushed much beyond a certain threshold value. This is the limiter action of the nonlinear process on the intensities. One can of course artificially raise the threshold to produce ω_{ss}, by an additive that has a specific strong absorption at ω_{ss}, but not at ω_s. In the same way as the vibrational wave at ω_v could be eliminated because of strong damping, so will strong damping at ω_{ss} eliminate this wave from the set of coupled equations. Since ω_{ss} is not created, higher order stokes lines will not be created either. In this case the solution (4-76) for just two coupled equations at ω_L and ω_s would appear to be applicable. In experiments carried out to date such suppression has not been supplied. Furthermore, if no resonator is used to select stokes waves of a particular direction, it is possible to generate copious amounts of radiation at the antistokes frequency $\omega_a = \omega_L + \omega_v$. This radiation discovered by Terhune[24] comes out in a forward cone mantle. Higher order antistokes frequencies at $\omega_L + \ell\omega_v$ are also observed. The coupling to these waves should also be taken into account.

4-6 COUPLING BETWEEN STOKES AND ANTISTOKES WAVES

In principle the problem is described by a set of coupled wave equations at the vibration frequency ω_v, the laser frequency ω_L, and all combination frequencies $\omega_L \pm \ell\omega_v$. These waves may still go in many directions. This problem must of course be made tractable by suitable approximations. In the first place the wave equation at ω_v can be eliminated because the optical phonons are heavily damped. Supply absorption for light at $\omega_L \pm 2\omega_v$. This will eliminate all higher order stokes and antistokes radiation for $\ell > 2$. Although a polarization at $\omega_L + 3\omega_v$ may be created by the combination frequency of antistokes and stokes $2\omega_a - \omega_s$, the coupling susceptibility for this process will not have a resonant denominator. For the same reason waves at harmonic frequencies, $2\omega_L$, etc. are eliminated from the discussion. The problem is thus reduced to coupled wave equations at three frequencies ω_L, ω_s, and ω_a. Even this problem is untractable. Therefore the laser field E_L will be treated as a fixed parameter. This approximation will, at least initially, represent the experimental situation fairly well, in which a laser beam of high intensity enters a Raman medium, crystal, liquid, or gas, at a plane boundary at $z = 0$.

The question how intensity at ω_s and ω_a is generated should then be answered by a solution of the coupled wave equations for $\mathbf{E}_s$ and $\mathbf{E}_a^*$. Little generality is lost by assuming the medium to be isotropic. Assume further that all waves are polarized in the same direction, so that the scalar Raman susceptibilities given by Eqs. (2-62) and (2-68) may be used. Stokes and antistokes polarization perpendicular to the incident beam could of course be treated in a similar manner. A real nonresonant part χ_{NR} has to be added to all Raman susceptibilities. The dispersion of this component may be ignored, because the excited electronic levels are distant compared to the separation $\omega_a - \omega_s$. The relationship between the complex quantities χ_a and χ_s has been discussed in Chapter 2.

The two coupled wave equations can now be written in the form,

$$\nabla^2 \mathbf{E}_s - \frac{\epsilon_s}{c^2}\frac{\partial^2 \mathbf{E}_s}{\partial t^2} = \frac{4\pi}{c^2}\frac{\partial^2}{\partial t^2}\left[(\chi_s + \chi_{NR})\,|E_L|^2\,\mathbf{E}_s + \left\{(\chi_a^* \chi_s)^{1/2} + \chi_{NR}\right\} E_L^2\, \mathbf{E}_a^*\right] \tag{4-77}$$

$$+\nabla^2 \mathbf{E}_a^* - \frac{\epsilon_a^*}{c^2}\frac{\partial^2 \mathbf{E}_a^*}{\partial t^2} = \frac{4\pi}{c^2}\frac{\partial^2}{\partial t^2}\left[\left\{(\chi_a^* \chi_s)^{1/2} + \chi_{NR}\right\} E_L^{*2}\, \mathbf{E}_s + (\chi_a^* + \chi_{NR})\,|E_L|^2\, \mathbf{E}_a^*\right] \tag{4-78}$$

The laser field is prescribed by $\mathbf{E}_L = 2\,\mathrm{Re}\,\{E_L \exp(i\mathbf{k}_L \cdot \mathbf{r} - i\omega_L t)\}$.

The symmetry of the boundary conditions is such that solutions are to be found in the form of plane waves with constant amplitudes in planes parallel to the boundary. Therefore try a solution of the form,

$$\mathbf{E}_s = E_s \exp(ik_{xs}x + ik_{ys}y)\exp(ik_{zs}z)\exp(-i\omega_s t) \tag{4-79}$$

$$\mathbf{E}_a^* = E_a^* \exp\{-i(2k_{xL} - k_{xs})x - i(2k_{yL} - k_{ys})y\} \times \exp - i(2k_{zL} - k_{zs})z \exp i(2\omega_L - \omega_s)t \tag{4-80}$$

Substitution of these expressions into the two wave equations reduces them to two homogeneous algebraic equations. The determinant can be put equal to zero. For each set of values for k_{xs}, k_{xy}, and ω_s a quartic equation for k_{zs} results. The negative imaginary part of a root k_{zs} signifies gain. The problem is to determine how this gain, if any, varies as a function of the tangential components, which determine the direction of the outgoing wave and the frequency ω_s. It must be remembered that χ_s and χ_a have a resonance behavior near $\omega_L - \omega_s = \omega_v$ and $\omega_a - \omega_L = \omega_v$, respectively.

Introduce the wave vectors in the linear medium in the absence of coupling, $E_L = 0$. Also allow for linear attenuation in the medium. Write

$$\epsilon_s \omega_s^2/c^2 \cong (k_s^0)^2 + 2i\alpha_s k_{sz}^m \tag{4-81}$$

$$\epsilon_a^* \omega_a^2/c^2 \cong (k_a^0)^2 - 2i\alpha_a k_{az}^m \tag{4-82}$$

The attenuation coefficients thus defined are a factor $(\cos\theta_s)^{-1}$ and $(\cos\theta_a)^{-1}$ larger than the attenuation coefficients measured along the direction of the wave. Furthermore introduce the z-components in the absence of coupling for the prescribed tangential components of the wave vectors.

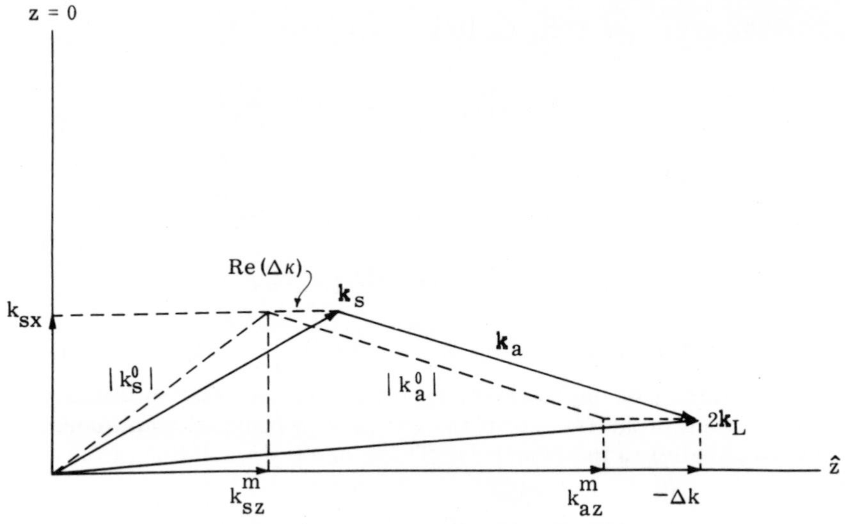

Figure 4-12. Geometrical relationships between the propagation vectors of stokes, antistokes, and laser radiation, with a momentum mismatch Δk in the direction normal to the boundary of the Raman cell.

$$k_{sz}^m = ((k_s^0)^2 - k_{sx}^2 - k_{sy}^2)^{1/2} \tag{4-83}$$

and

$$k_{az}^m = ((k_a^0)^2 - k_{ax}^2 - k_{ay}^2)^{1/2} = \{(k_a^0)^2 - (2k_{Lx} - k_{sx})^2 - (2k_{Ly} - k_{sy})^2\}^{1/2} \tag{4-84}$$

The phase mismatch in the z-direction for the wave vectors in the absence of coupling is, as shown in Figure 4-12,

$$-\Delta k = 2k_{Lz} - k_{sz}^m - k_{az}^m \tag{4-85}$$

We anticipate solutions for k_{sz} which differ only slightly from k_{sz}^0 and introduce the new variable

$$\Delta\kappa = k_{sz} - k_{sz}^m \tag{4-86}$$

When the determinant of the coupled wave problem is put equal to zero, the following quartic equations for $\Delta\kappa$ results,

$$\begin{vmatrix} 2k_{sz}^{m}(\Delta\kappa - i\alpha_s) + (\Delta\kappa)^2 - \frac{4\pi\omega_s^2}{c^2}(\chi_s + \chi_{NR})|E_L|^2 & -\frac{4\pi}{c^2}\omega_s^2\{(\chi_a^*\chi_s)^{1/2} + \chi_{NR}\}E_L^2 \\ -\frac{4\pi\omega_a^2}{c^2}\{(\chi_a^*\chi_s)^{1/2} + \chi_{NR}\}E_L^{*2} & \begin{matrix} -2k_{az}^{m}(\Delta\kappa + \Delta k - i\alpha_s) \\ + (\Delta\kappa + \Delta k)^2 - \frac{4\pi\omega_a^2}{c^2} \\ \times (\chi_a^* + \chi_{NR})|E_L|^2 \end{matrix} \end{vmatrix} = 0 \tag{4-87}$$

First consider the case $\Delta k = 0$. This corresponds to perfect matching of the wave vectors for the linear medium, $2\mathbf{k}_L - \mathbf{k}_s - \mathbf{k}_a = 0$. For equal linear loss $\alpha_a = \alpha_s = \alpha_{sa}$ and ignoring small dispersion effects in the Raman susceptibility $\chi_a^* \approx \chi_s$, inspection of the quartic determinant equation shows that it has a double root $\Delta\kappa = i\alpha_{sa}$. This means that for the direction or the value of the tangential component for which the momentum matching condition is fulfilled, neither stokes nor antistokes radiation is generated. This rather unexpected result comes about because the positive work done on one part of the normal mode near ω_s is exactly compensated by the negative work done on the other part near ω_a. This situation is well known in parametric amplifier theory, where no gain can be obtained at $\omega_s = \omega_L - \omega_i$ if the other side band at $\omega_L + \omega_i$ is not suppressed.

To make further progress the quartic is reduced to a quadratic equation by dropping the terms $(\Delta\kappa)^2$ and $(\Delta\kappa + \Delta k)^2$. This is a good approximation because $|\Delta k| \ll k_{az}$ and because the interest is in solutions for which $|\Delta\kappa| \ll k_{sz}$. The two roots that are abandoned correspond to backward traveling waves. The quadratic equation that remains can be put in the form,

$$(\Delta\kappa - \lambda_{ss})(\Delta\kappa - \lambda_{aa}) - \lambda_{as}\lambda_{sa} = 0 \tag{4-88}$$

with

$$\lambda_{ss} = i\alpha_s + (2\pi\omega_s^2/c^2k_{sz}^{m})(\chi_s + \chi_{NR})|E_L|^2$$

$$\lambda_{aa} = -\Delta k + i\alpha_a - (2\pi\omega_a^2/c^2k_{az}^{m})(\chi_a^* + \chi_{NR})|E_L|^2$$

$$\lambda_{as}\lambda_{sa} = -(4\pi^2\omega_s^2\omega_a^2/c^4k_{sz}^{m}k_{az}^{m})\{(\chi_a^*\chi_s)^{1/2} + \chi_{NR}\}^2|E_L|^4$$

Precisely the same quadratic equation could have been obtained by starting from the coupled amplitude equations.[22]

The two roots of Eq.(4-88) can of course be written down quite generally. In practice, the situation in which the attenuation coefficients are equal, $\alpha_s = \alpha_a = \alpha_{as}$, and where Raman dispersion effects may be ignored, $\chi_s \omega_s^2 / k_{sz}^m = \chi_a^* \omega_a^2 / k_{az}^m$, is of most interest and gives some algebraic simplification. The two roots are then

$$\Delta\kappa = i\alpha_{sa} - \tfrac{1}{2}\Delta k \pm [\tfrac{1}{4}\Delta k^2 + (2\pi\omega_s^2/c^2 k_{sz}^m)(\chi_s + \chi_{NR})|E_L|^2 \Delta k]^{1/2} \tag{4-89}$$

For $\Delta k = 0$ both roots are $i\alpha_{sa}$, signifying loss in both modes. If Δk is large, the square root can be expanded and the roots become

$$\begin{aligned}\Delta\kappa_s &= i\alpha_{sa} + (2\pi\omega_s^2/c^2 k_{sz}^m)(\chi_s + \chi_{NR})|E_L|^2 \\ &\quad - (2\pi\omega_s^2/c^2 k_{sz}^m)^2(\chi_s + \chi_{NR})^2 |E_L|^4 \Delta k^{-1}\end{aligned} \tag{4-90}$$

$$\Delta\kappa_a = i\alpha_{sa} - \Delta k - (2\pi\omega_s^2/c^2 k_{sz}^m)(\chi_s + \chi_{NR}) |E_L|^2 + \cdots \tag{4-91}$$

The first root $\Delta\kappa_s$ clearly corresponds to a wave with almost pure stokes character. The imaginary part of χ_s is negative and the wave has gain if α_{sa} is not too large. In the limit $\Delta k \gg (2\pi\omega_s^2/c^2 k_{sz}) \times \chi_s |E_L|^2$, this is precisely the same stokes gain, calculated previously in Eq.(4-68). The other root corresponds to the almost pure antistokes wave which is always attenuated.

If the eigenvalue $\Delta\kappa_s$ given by Eq. (4-90) is reinserted into the coupled equations, one finds that the partial antistokes character, admixed by the laser field to the stokes wave, is given by

$$|E_{as}|^2/|E_s|^2 = (2\pi\omega_s^2/c^2 k_{sz}^m)^2 |\chi_s + \chi_{NR}|^2 |E_L|^4 \Delta k^{-2} \ll 1 \tag{4-92}$$

If Δk is chosen smaller, this approximation loses its validity. In the vicinity of the phase-matched direction the coupling between the stokes and antistokes wave becomes quite strong. The detailed behavior of the coupling is given by Eq. (4-89). Since the square root of a complex quantity is involved, the behavior of the gain corresponding to the imaginary part of this square root is not easily visualized. For each value of Δk, i.e., for each value of the transverse momentum $(k_{sx}^2 + k_{sy}^2)^{1/2}$, the gain coefficient can be found as a function of the frequency ω_s. The resonant part χ_s is of course a rapidly changing function of ω_s. Its real and imaginary part have the well-known behavor of a linear susceptibility near resonance as shown in in Figure 2-2.

The gain coefficient as a function of Δk is plotted in Figure 4-13.

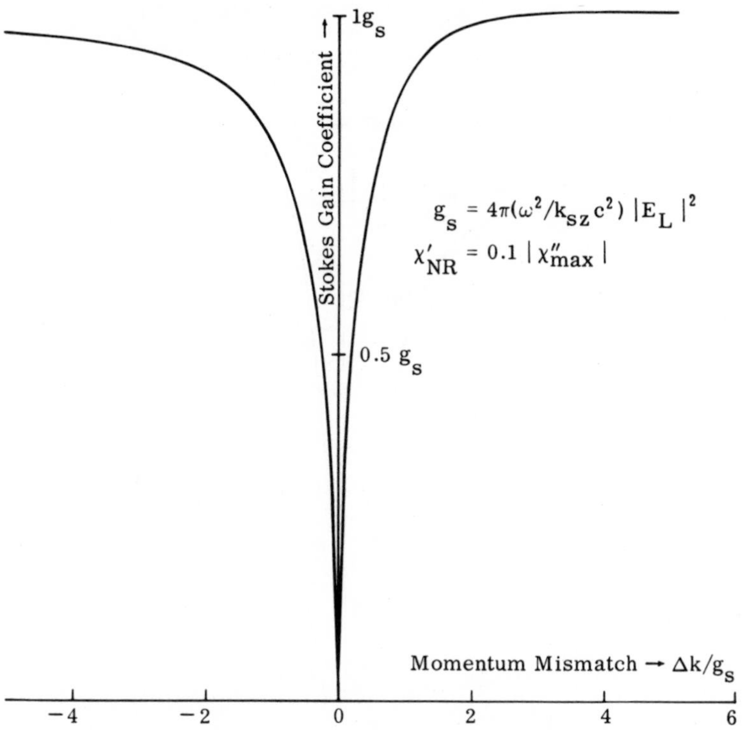

Figure 4-13. The gain coefficient of the coupled stokes-antistokes wave as a function of the linear momentum mismatch Δk in the z-direction. The curve is normalized to the stokes gain g_s far from the matched direction.

It drops sharply to zero in the exact phase-matched direction. The asymmetry is caused by the nonresonant part. The choice $\chi_{NR} > 0$ and $|\chi_{NR}/\chi_s''^{max}| = 0.1$ has been made in computing the curve. If the gain coefficient is multiplied by the partial antistokes character of the wave, a machine calculation gives the results of Figure 4-14. For each value of Δk, the frequency off-set $\Delta\omega$ has been chosen so as to maximize the gain. At the maximum antistokes gain one finds $\Delta\omega \approx 0.3\Gamma$ in this example. In general one may expect frequencies in a range of the order of $0.5\Gamma/2\pi$ to be amplified.

The figure is normalized by expressing Δk in units of stokes power gain per centimeter from Eq. (4-68), $g_s = 4\pi\omega_s^2/c^2 k_{sz}^m \times |\chi_s''|E_L|^2$. In these units the maximum gain occurs for $\Delta k/g_s = 2$. The direction of the wave corresponding to a given Δk is found

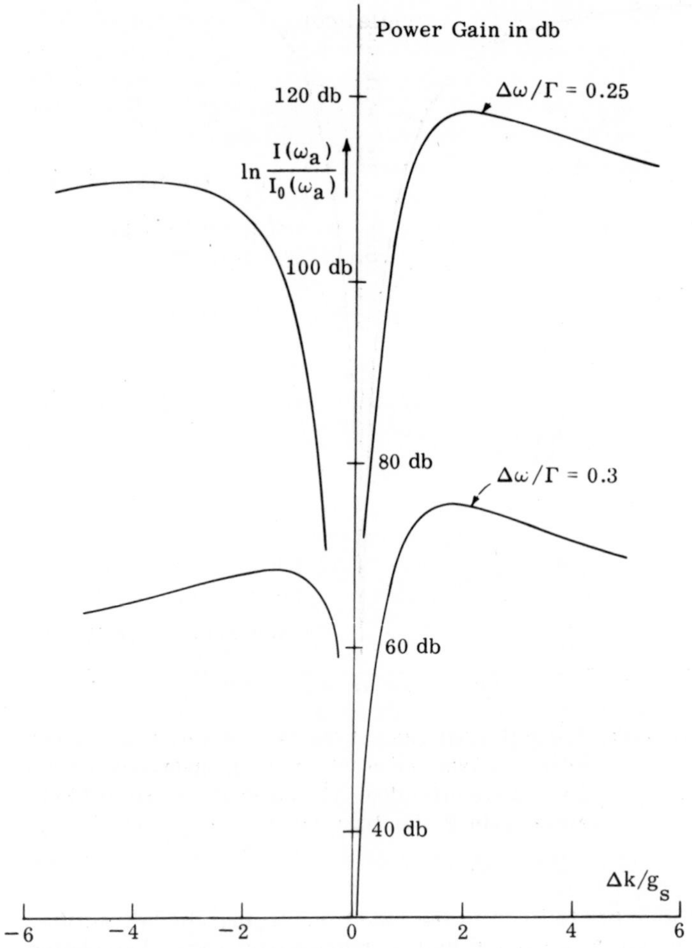

Figure 4-14. The antistokes gain as a function of the normalized linear momentum mismatch in the z-direction $\Delta k/g_s$, with $g_s = 4\pi(\omega^2/c^2 k_{sz}^m)|\chi_s''|\ |E_L|^2$. For each Δk the direction of the radiation outside the Raman cell can be found from Figure 4-12. For each direction of observation there is a frequency off-set $\Delta\omega = \omega - \omega_L + \omega_v$ for maximum gain. The asymmetry around $\Delta k = 0$ is due to the nonresonant part, $|\chi'_{NR}/\chi''_s \text{ max}| = 0.1$.

from Eqs. (4-83), (4-84), and (4-85). The geometry is shown in Figure 4-12. For a given tangential component, given k_s^0 and k_a^0 and given laser beam $\mathbf{k}_L$, the whole figure can be constructed. A given Δk is thus related to the value of transverse momentum. The deviation of the direction of maximum gain from the direction θ_0 for perfect momentum matching is thus determined, since the tangential components at the boundary in turn prescribe the direction of the beam outside the Raman medium.

The threshold for the generation of waves is a function of the direction. In Raman cells, long compared to the diameter of the laser beam, geometrical factors will play a dominant role. But even for the idealized case of plane waves of infinite lateral extent, there is a very sharp rise in the threshold near the phase-matched direction. In this exact direction the threshold is infinite. There should therefore be a dark ring in the directional intensity pattern of the Stokes radiation, given by the phase-matched direction. There should be a bright ring—or perhaps two, if χ_{NR} is very small—for the antistokes radiation in a direction with an off-set from the phase-matched direction. A small fraction of the intensity in any direction should be at the antistokes frequency. The ratio of the intensity of stokes to antistokes radiation is given by Eq. (4-92). It should be emphasized that these results are derived for the condition that radiation at $\omega_L \pm \ell\omega_v$ is suppressed for $\ell \geq 2$, and that the laser field can be considered as a constant parameter and is not depleted. Neither of these conditions are satisfied in the usual experiments on coherent Raman generation. A comparison with the experimental results is postponed till the next chapter.

If is of interest to consider also the solution of the Raman type amplifier. In this case a light beam, either at ω_s or ω_a is incident, in addition to the laser beam. The direction of the incident beam, say at the prescribed frequency ω_s,

$$E_s^{inc} \exp(i\mathbf{k}_s^i \cdot \mathbf{r} - i\omega_s t)$$

with respect to the boundary $z = 0$ determines the tangential components at the start. Therefore, Δk is determined. One must now take the appropriate linear combination of the two normal modes, corresponding to this Δk and ω_s, to match the boundary condition.

Denote the two roots by $\Delta\kappa_1$ and $\Delta\kappa_2$ with corresponding eigenfunctions,

$$A_1\{c_{1s} \exp i(k_{sz}^m + \Delta\kappa_1)\, z + c_{1a}^* \exp -i(k_{az}^m - \Delta k - \Delta\kappa_1)z\} \tag{4-93}$$

$$A_2\{c_{2s}\exp(ik^m_{sz} + \Delta\kappa_2)z + c^*_{2a}\exp - i(k^m_{az} - \Delta k - \Delta\kappa_2)z\} \tag{4-94}$$

The complex "eigenvectors" are normalized, $|c_{1s}|^2 + |c_{1a}|^2 = 1$, $|c_{2s}|^2 + |c_{2a}|^2 = 1$, with $c_{1s}/c^*_{1a} = E_{s,1}/E^*_{a,1} = f(\Delta\kappa_1)$ and $c_{2s}/c^*_{2a} = E_{s,2}/E^*_{a,2} = f(\Delta\kappa_2)$. The x,y,t dependence of the functions has been omitted because they are completely prescribed by Eq. (4-80).

The amplitudes of the eigenmodes are now determined by the boundary condition at $z = 0$. The complete solution would, in addition, involve two reflected waves, at ω_s and ω_a, respectively. This gives a total of four amplitudes to be determined from the continuity of the tangential components of E and H, at ω_s and at ω_a separately. From the considerations of section 4-1 it is known that the reflected amplitudes will be rather small, and a tentative approximation will be obtained by taking the following two equations to determine A_1 and A_2,

$$c_{1s}A_1 + c_{2s}A_2 = E^{inc}_s \quad \text{and} \quad c_{1a}A_1 + c_{2a}A_2 = 0 \tag{4-95}$$

The second boundary condition will have to be modified in a way that will shortly become apparent. When these amplitudes are reinserted into Eqs. (4-93) and (4-94), both the stokes and the antistokes component are seen to be a linear combination of two exponentials. Except for the exact phase-matched direction, one of these will be a wave with exponential gain. For sufficiently long distance z this is the only term that has to be retained. The direction in which the stokes energy emerges on the other side of the plane-parallel Raman cell is of course parallel to incident beam. The antistokes light emerges in general at a different angle determined by its tangential components.

For very small values of z, the antistokes amplitude is proportional to z in agreement with the general considerations of section 4-1. Expansion of the exponentials gives for $|\Delta\kappa_s| \ll 1$,

$$E_a = i(c_{1a}A_1\Delta\kappa_1 + c_{2a}A_2\Delta\kappa_2)z\exp(ik^0_{az}z)$$

In the exact phase-matched direction the two roots are both exponentially decaying. They are equal for the conditions under which Eq. (4-89) was derived. In that degenerate case the Eqs. (4-93) and (4-94) are not linearly independent, and E_a according to Eq. (4-96) would become identically zero. The correct solution then has the form $\Delta\kappa_1 = \Delta\kappa_2 = i\alpha_{sa}$

$$\lim_{\Delta\kappa_1 - \Delta\kappa_2 \to 0} E_a = -A_2 z\exp(ik^0_{az}z - \alpha_{sa}z) \tag{4-97}$$

The value of dE_a/dz for small z is in this case best determined from the considerations of section 4-1. For small z the solution has to match the expression given by Eq. (4-12). It is in fact possible also for $\Delta\kappa_1 \neq \Delta\kappa_2$, to replace the initial condition $c_{1a}A_1 + c_{2a}A_2 = 0$ of Eq. (4-95) by the condition that $i(c_{1a}A_1\,\Delta\kappa_1 + c_{2a}A_2\,\Delta\kappa_2)z$ represents the correct initial parametric growth given by Eq. (4-12).

This completes the discussion of coupled wave problems. The arguments are of course not restricted to light waves and acoustical waves. The interaction between light waves and spin waves could be treated along similar lines. In fact, the parametric interaction between coupled spin waves was discussed before the birth of nonlinear optics by Suhl and Tien.[25] The interaction between electromagnetic waves and electron density and velocity waves in traveling wave tubes has also many of the same properties.[26] The reader familiar with parametric amplifier theory will have recognized many familiar features. The most interesting new aspects are inherent in the three-dimensional character and the dispersive properties of the light waves. This has led to a generalization of many familiar laws of optics to nonlinear situations. The theoretical results will now be compared with the experimental findings.

REFERENCES

1. N. Bloembergen and P. S. Pershan, *Phys. Rev.*, **128**, 606 (1962); this paper will henceforth be referred to and is reproduced as Appendix 2.
2. J. Ducuing and N. Bloembergen, *Phys. Rev. Letters*, **10**, 474 (1963).
3. N. Bloembergen and J. Ducuing, *Physics Letters*, **6**, 5 (1963).
4. R. Kronig and J. I. Boukema, *Proc. Royal Neth. Acad.* (Amsterdam), **66**B, 8 (1963).
5. R. W. Terhune, P. D. Maker, and C. M. Savage, *App. Phys. Letters*, **2**, 54 (1963).
6. N. S. Shiren, *App. Phys. Letters*, **4**, 82 (1964).
7. J. M. Ziman, *Electrons and Phonons*, Clarendon Press, Oxford, 1960, Chap. 3.
8. M. Born and K. Huang, *Dynamical Theory of Crystal Lattices*, Clarendon Press, Oxford, 1954.
9. R. H. Kingston, *Proc. IRE*, **50**, 472 (1962).
10. N. M. Kroll, *Phys. Rev.*, **127**, 1207 (1962).
11. N. M. Kroll, *J. App. Phys.*, to be published.
12. R. Chiao, C. H. Townes, and B. Stoicheff, *Phys. Rev. Letters*, **12**, 592 (1964). R Chiao, E. Garmire and C. H. Townes, *Rendiconti S.I.F. Corso 22*, August 1963, Academic Press, New York, 1964.

13. R. W. Hellwarth, *Phys. Rev.*, **130**, 1850 (1963).
14. H. J. Zeiger and P. E. Tannenwald, *Proceedings 3rd International Conference on Quantum Electronics*, Paris, February 1963, ed. P. Grivet and N. Bloembergen, Columbia University Press, New York, 1964, p. 1589.
15. N. Bloembergen, *ibidem*, p. 1501.
16. R. Loudon, *Proc. Phys. Soc.* (London), A **82**, 393 (1963), *Proc. Roy. Soc.*, A **275**, 218 (1963).
17. R. W. Terhune, *Solid State Design*, **4**, 38, November 1963.
18. E. Garmire, E. Pandarese, and C. H. Townes, *Phys. Rev. Letters*, **11**, 160 (1963).
19. A. Javan, *Rendiconti S.I.F. Corso 22*, August 1963, Academic Press, New York, 1964.
20. R. W. Hellwarth, *App. Optics*, **2**, 847 (1963). R. W. Hellwarth, *Current Sci.* (India), **33**, 129 (1964).
21. H. J. Zeiger, P. E. Tannenwald, S. Kern, and R. Herendeen, *Phys. Rev. Letters*, **11**, 419 (1963).
22. N. Bloembergen and Y. R. Shen, *Phys. Rev. Letters*, **12**, 504 (1964).
23. G. Eckhardt, R. W. Hellwarth, F. J. McClung, S. E. Schwarz, and D. Weiner, *Phys. Rev. Letters*, **9**, 455 (1962).
24. R. W. Terhune, *Bull. Am. Phys. Soc.*, II, **8**, 359 (1963).
25. P. K. Tien and H. Suhl, *Proc. IRE* **46**,700 (1958).
26. J. R. Pierce, *Traveling-Wave Tubes*, Van Nostrand, New York, 1950.

5

EXPERIMENTAL RESULTS

5-1 EXPERIMENTAL VERIFICATION OF THE LAWS OF NONLINEAR TRANSMISSION AND REFLECTION

All experiments performed to date on parametric generation of light are described quite well by the equations of sections 4-1 and 4-2. The most elaborate tests come from the second-harmonic generation, SHG, of light in piezoelectric crystals. Franken's first experiment[1] was performed with a ruby laser beam at 6943 Å, focused to a spot about 1 mm in diameter in a quartz crystal. The laser pulse of about 3 joules energy lasted about 10^{-3} sec. Radiation from the quartz was detected with a quartz spectrograph on red insensitive film. It was estimated that about $1{:}10^{11}$ photons were converted to ultraviolet quanta at 3471 Å. Later experiments with gratings have confirmed that the frequency generated is twice the frequency of the ruby laser with the experimental accuracy of one part in 10^5. The first order diffraction in the UV is compared with the red diffraction in second order in the same exposure.

The early experimental results, mentioned in Chapter 1 are well described by the Eqs. (4-12) and (4-13). In particular the results of the transmission through a plane-parallel slab of varying optical thickness, an example of which was shown in Figure 1-2 can be fitted by Eq. (4-13). The dispersion of momentum mismatch $\epsilon^{1/2}(2\omega) - \epsilon^{1/2}(\omega)$ may be determined from the distance between maxima, whereas the height of the maxima is a measure for the magnitude of the nonlinear susceptibility. The basic original geometry, shown in Figure 1-1, is still used for the determination of the nonlinear susceptibilities with the aid of Eq. (4-13).

Giordmaine[2] and Maker[3] et al., simultaneously pointed out that in a negative uniaxial crystal with a sufficiently large birefringence, the ordinary wave at the fundamental frequency has the same propagation constant as the extraordinary ray at twice the frequency for a particular direction of the light waves with respect to the optic axis. For $\lambda = 6940$ Å in potassium dihydrogen phosphate, KDP, this occurs when the ruby light beam, polarized perpendicular to the optic axis, makes an angle of $49.8 \pm 0.5°$ with the optic axis. The intensity of SHG can thus be increased by many orders of magnitude. For exact phase matching it is proportional to the square of the thickness of the crystal as follows from Eq. (4-12). It is, however, a difficult problem to retain phase matching over long distances, because the incident ruby laser beam contains many modes. It is then advantageous to focus the laser beam. The increased intensity at the focus shortens the interaction length, given by Eq. (4-35). Terhune, Maker, and Savage[4] used a 25 cm focal distance lens to focus the output of a "giant pulse" or "Q-switched" laser, which puts out about 0.1 joule in 20×10^{-9} sec, in a crystal of $(NH_4)_2 H_2 PO_4$ (ADP). They were able to convert more than twenty per cent of the red light into second harmonic intensity. They observed the depletion of the pump, described by Eq. (4-34). It is interesting to note that KDP could not be used in this experiment. The ultraviolet absorption edge was near the third harmonic for KDP, and triple photon absorption damaged this crystal, whereas the same intensity in ADP, with an absorption band further in the ultraviolet, did no harm.

Geometrical Considerations

The ideal focus of a diffraction limited system is described by a numerical aperture or f/number,

$$f = \beta^{-1} = n\ell_f/d$$

where ℓ_f is the focal length in vacuum, d the beam diameter, n the index of refraction in the medium where the focus occurs, and β is the full apex angle of the light cone emanating from the focal point. The theory of the ideal focus for $f \gg 1$ has been worked out in detail in diffraction theory.[5] The region of high intensity near the focus can be described approximately by a plane wave in a cylindrical region of length

$$\xi = 8f^2\lambda_0/n \qquad (5\text{-}1)$$

and diameter

$$\delta = 2\lambda_0 f/n \qquad (5\text{-}2)$$

The wave length in this region is compressed by a factor $\{1 - (4f)^{-2}\}$.

The power flux density in the focus is increased by a factor $(d/\delta)^2$. This increase is clearly not achieved in the actual experiment, because the light in the ruby beam is not diffraction limited. Assume an aperture of 4×10^{-3} radians for the unfocused beam. The diameter of the light beam near the focus for $\ell_f = 25$ cm is 0.1 cm. This is ten times smaller than the diameter of the unfocused beam which is about 1 cm across. The power flux density in the focus is approximately 100 times larger than in the unfocused beam or about 500 megawatts/cm². The interaction length according to Eq. (4-35) is then of the order of 2mm. This is about twice the length of the focal region ξ. Eq. (4-34) would predict about the observed percentage of conversion in a length ξ. There are several reasons why this equation does not apply. The detailed nature of the mode structure in the laser beam and in the focus is not known. The intensity is not uniform in the cross section. There are pencil rays in the beam which are nearly completely converted. Other pencil rays notably near the circumference of the beam will undergo little or no conversion. This situation is aggravated by the anisotropic nature of the medium. Not all directions in a beam of finite aperature can have perfect phase matching.

Kleinman[6] has given a discussion of SHG in uniaxial crystals, when beams of a finite aperture rather than plane waves of infinite cross section are considered. Assume a uniform intensity distribution of the fundamental in a small interval $\Delta\theta$ around the exact phase-matched direction θ_0. The angular distribution of SHG in the parallel slab of thickness z is proportional to the function $\sin^2\{(n(\theta) - n(\theta_0))\,\omega c^{-1} z\}\{n(\theta) - n(\theta_0)\}^{-2}$. The index of refraction of the extraordinary ray is a function of the angle $n(\theta)$. The total second-harmonic intensity, averaged over all directions, is proportional to,

$$\frac{1}{\Delta\theta}\int_{-\frac{1}{2}\Delta\theta}^{\frac{1}{2}\Delta\theta} \frac{\sin^2(\partial n/\partial\theta_0)\,\omega c^{-1} z\theta}{(\partial n/\partial\theta_0)^2\,\theta^2}\, d\theta \qquad (5\text{-}3)$$

For $(\partial n/\partial\theta_0)\,\omega c^{-1} z\Delta\theta \ll 1$, this integral may be approximated by $\omega^2 c^{-2} z^2$. For $(\partial n/\partial\theta_0)\,\omega c^{-1} z\Delta\theta \gg 1$, the integral may be developed as

$$\omega c^{-1} z\left(\frac{\partial n}{\partial\theta_0}\,\Delta\theta\right)^{-1}\int_{-\infty}^{+\infty} u^{-2}\sin^2 u\,du = \pi\omega c^{-1} z\,\{(\partial n/\partial\theta_0)\,\Delta\theta\}^{-1}$$

An average effective coherence length for a beam with angular spread $\Delta\theta$ may be defined by,

$$z = \ell'_{coh} = \{(\partial n/\partial\theta_0)\,\omega c^{-1}\,\Delta\theta\}^{-1} \tag{5-4}$$

For KDP at 3470 Å, $n^{-2} = (1.534)^{-2}\cos^2\theta + (1.487)^{-2}\sin^2\theta$. The matching angle is at $\theta_0 = 49.9°$. Therefore one obtains with $\partial n/\partial\theta_0 \approx 0.03$ and a typical ruby beam spread $\Delta\theta = 4 \times 10^{-3}$ radians, an effective length $\ell'_{coh} = 0.1$ cm. For $z > \ell'_{coh}$ the second-harmonic intensity will only increase proportional to z, and not as z^2. Fortunately, ℓ'_{coh} is of the same magnitude as the focal depth and the interaction length, so that the previous considerations are not invalidated.

Giordmaine[2] has also discussed SHG by two fundamental rays that do not travel in the same direction. Phase matching in a momentum triangle is possible between two ordinary rays and one extraordinary harmonic ray in directions near θ_0. Interesting intensity patterns are observed which confirm the theoretical considerations of phase matching.

Reflected Harmonic Waves

Several workers have observed the second-harmonic generation in transmission through a slab which strongly absorbs this frequency, but transmits the fundamental wave. This effect is described by the polarization wave of Eq. (4-15).

When the medium absorbs both the fundamental and the harmonic frequency, SHG is only possible in reflection. This effect was first detected by Ducuing.[7] The direction of the reflected harmonic rays has already been discussed in section 4-1. The polarization characteristics of the SHG in reflection from GaAs were measured with the experimental arrangement shown in Figure 5-1. The third-rank tensor $\chi_{ijk}(2\omega = \omega + \omega)$ in $\bar{4}3m$ symmetry has the following nonvanishing elements $\chi_{xyz} = \chi_{zxy} = \chi_{yzx} = \frac{1}{2}\chi_{14}$. The incident laser beam is polarized normal to the plane of incidence. The components of nonlinear polarization with respect to the cubic crystallographic axes are,

$$\begin{aligned} P_x(2\omega) &= \chi_{14} E_y E_z \\ P_y(2\omega) &= \chi_{14} E_z E_x \\ P_z(2\omega) &= \chi_{14} E_x E_y \end{aligned} \tag{5-6}$$

When the crystal, which serves as a "nonlinear mirror," is rotated around its own normal with crystallographic direction $[00\bar{1}]$, the fundamental polarization can be made parallel to the [100], [111], and [011] directions, respectively. In the first case, $E_y = E_z = 0$, and

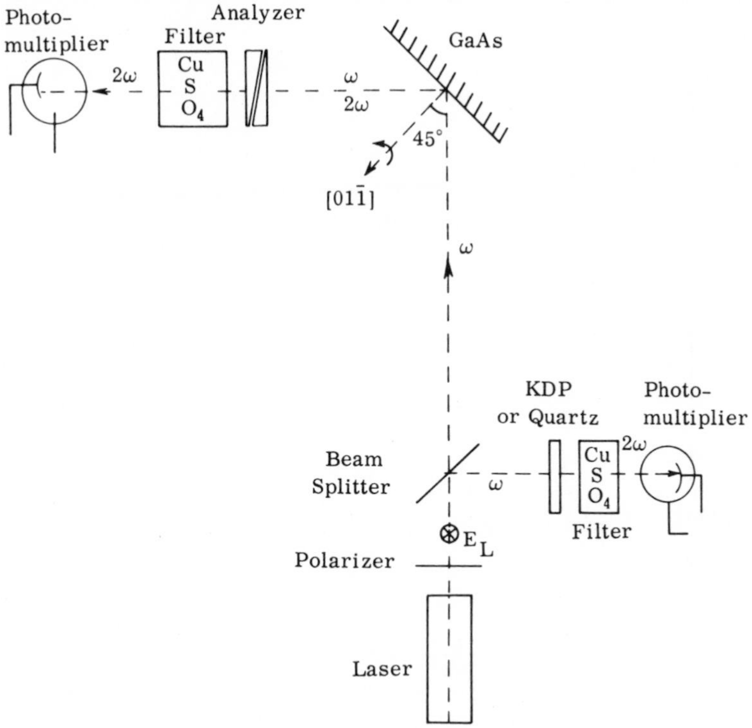

Figure 5-1. Experimental arrangement for the measurement of the nonlinear susceptibility of GaAs in reflection.

there is no second-harmonic polarization induced. For $E(\omega)$ parallel to [111], $E_x(\omega) = E_y(\omega) = E_z(\omega)$ and Eqs. (5-6) immediately show that $P(2\omega)$ is also parallel to [111]. Therefore, the polarization of the reflected harmonic is also perpendicular to the plane of incidence. For $E_x = 0$, $E_y = E_z$, the harmonic polarization has only an x-component. $\mathbf{P}(2\omega)$ is then at right angles to the polarization of the incident laser beam. The polarization of the reflected harmonic ray lies in the plane of reflection. These features are strikingly confirmed by the experimental results shown in Figures 5-2 and 5-3. The experimental points for the intensity of the parallel and perpendicular component of reflected second-harmonic intensity are compared with the theoretical curves for an arbitrary angle ψ between the direction of the fundamental polarization and the crystallographic x-axis. For the geometry shown in Figure 5-1 the angular dependence of the intensity for the two polarizations of the reflected harmonic is given by,

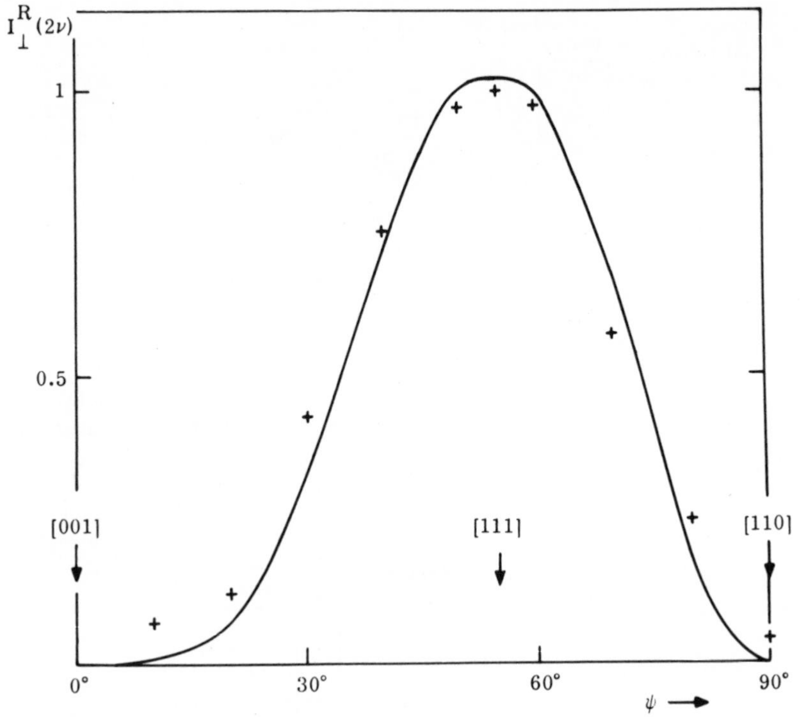

Figure 5-2. Intensity of the second-harmonic reflected beam from GaAs as a function of the angle between the plane of incidence and the crystallographic [100] axis. Both the fundamental and second harmonic are polarized normal to the plane of incidence. The beam is calibrated with a nonlinear process of the same order, where KDP or quartz is used as a standard.

$$I_{\perp}(2\omega) \propto P_{\perp}^2(2\omega) \propto \sin^4\psi \cos^2\psi$$
$$I_{\parallel}(2\omega) \propto P_{\parallel}^2(2\omega) \propto \sin^2\psi (1 - 3\cos^2\psi)^2 \qquad (5\text{-}7)$$

A GaAs slab is transparent for a laser beam at $\lambda = 1.06$ microns obtainable from a neodymium glass laser. In this case the ratio of the reflected to transmitted intensity in SHG can be determined. According to Eqs. (4-8) and (4-15) this should be independent of the nonlinear susceptibility of the material. These equations indeed correctly yield the ratio observed by Ducuing. This may be considered a check on the generalized Fresnel law for the harmonic reflected intensity.

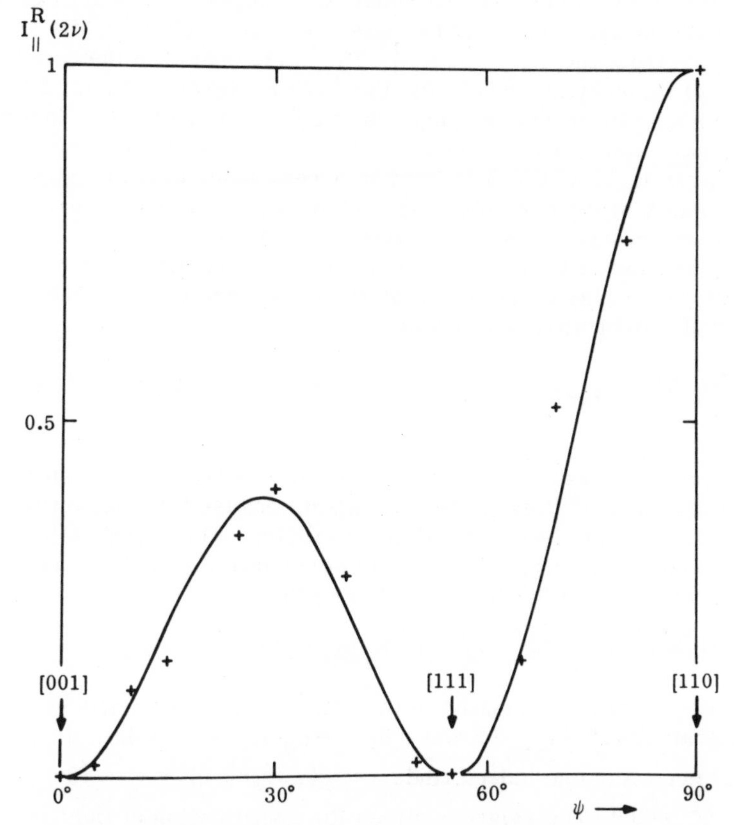

Figure 5-3. Intensity of the second harmonic reflected beam from GaAs, as a function of the angle between the plane of incidence and the crystallographic [100] axis. The fundamental ray is polarized normal to the plane of incidence, the second-harmonic polarization lies in this plane.

Generation of Sum and Difference Frequencies

Franken and coworkers[8] have first obtained the sum frequency of two light beams at different frequencies. They used two ruby lasers, operating at different temperatures, so that their wave lengths were about 10 Å apart. They indeed observed three ultraviolet frequencies, two second harmonics and the sum frequency in between. Smith and Braslau[9] observed the difference frequency, obtained by beating a ruby laser beam with the 3115 Å line of a mercury discharge. This latter source is of course incoherent and its light should be considered as a superposition of a large number of plane waves. Each of these beats with the laser beam and creates an observable intensity at the difference frequency, $\lambda = 5650$ Å. They also observed the sum frequency of the ruby laser line and the 5470 Å mercury line at 3056 Å The sum and difference frequencies of ruby and neodymium laser beams have also been obtained. Terhune has obtained a frequency 3ω by first doubling the frequency of a ruby laser beam in KDP and subsequently beating the beam at 2ω and a portion of the beam at ω in another crystal in a suitable phase-matched orientation.

Franken and coworkers[10] also observed the rectification of light in KDP, which may be regarded as the difference beat at zero frequency of a light wave with itself,

$$P_z(0) = \chi_{zxy}(0 = \omega - \omega)\{E_x(\omega)E_y^*(-\omega) + E_x^*(-\omega)E_y(\omega)\} \tag{5-8}$$

There is considerable uncertainty, perhaps as large as a factor of two, in the measurement of this constant because the interaction volume of the light beam is difficult to define. This observation provides an interesting check on the permutation symmetry relation by comparison with the linear dc Kerr effect.

$$P_x(\omega) = \chi_{xyz}(\omega = \omega + 0)\, E_y(\omega)\, E_z(0) \tag{5-9}$$

This susceptibility is related to the constant r_{36}, normally quoted in the literature, by $+\chi_{xyz}(\omega = \omega + 0) = -n_o^4 r_{36}/4\pi$. Within the experimental accuracy one indeed finds $\chi_{xyz}(\omega = \omega + 0) = \chi_{zxy}(0 = \omega - \omega) = 13 \times 10^{-8}$ esu.† The relative values for rectification of light in KH_2PO_4 and KD_2PO_4 are in excellent agreement with the corresponding ratio for the linear electro-optic constant.

When the dc field in Eq. (5-9) is replaced by an ac field at radio or microwave frequencies, one obtains side bands on the light carrier.

$$\begin{aligned} P_y(\omega_L + \omega_m) &= \chi(\omega = \omega_L + \omega_m)\, E_x(\omega_L)\, E_z(\omega_m) \\ P_y(\omega_L - \omega_m) &= \chi(\omega = \omega_L - \omega_m)\, E_x(\omega_L)\, E_z^*(-\omega_m) \end{aligned} \tag{5-10}$$

These side bands of light have been resolved with a grating spectrometer for $\omega_m/2\pi = 15$ kmc/sec. Kaminow[11] and many other authors have built practical microwave modulators for light. Many interesting

†Note that with our definition of the ac amplitudes, the value of $\chi_{zxy}(0 = \omega - \omega)$ is four times as large as the value of Ward and Franken.[10] Because we distinguish positive and negative frequencies, the symmetry relation does not contain a factor 4 in our case.

phase-matching devices are possible in the combination of guided microwaves and light waves. The theory of plane waves in free space can of course be extended to other wave modes. The inverse process, in which two light frequencies beat together to produce a polarization at a microwave frequency which radiates microwave power has been observed by Niebuhr.[12] Two axial modes of a ruby laser, spaced in frequency by 2.964 kmc/sec, were mixed in a quartz crystal, which was simultaneously part of the laser resonator and a microwave resonant cavity.

The conclusion of this section is that the equations presented in sections (4-1) and (4-2) give an accurate description of the wide variety of observed parametric phenomena in the optical nonlinearities of lowest order. Conversely these equations may be used to derive numerical values for the nonlinear susceptibilities from the experimental observations.

5-2 ABSOLUTE DETERMINATION OF A NONLINEAR SUSCEPTIBILITY

The difficulty in obtaining a reliable value of the nonlinear constant from the observed second harmonic intensity is connected with the multimode structure of the high-power solid state lasers. The intensity distribution in each cross section along the path of the beam is not precisely known. The pulsed nature of the signal adds a further difficulty. The second-harmonic intensity is not proportional to the square of integrated fundamental intensity, but to the integrated (fundamental intensity).[2] This difficulty will be analyzed more fully in the next section.

Light beams from gas lasers have a well-defined geometry and the intensity is constant in time. Ashkin, Boyd, and Dziedzic[13] used a helium neon laser at 1.1526 micron wave length, in the fundamental mode (00q) with a radial intensity distribution $I_1 = I_0 \exp(-2r^2/a^2)$, with spot size $a = 0.21$ cm. The half beam angle of the diffraction limited beam is only 3×10^{-5} radian. This is important for the measurement of the nonlinearity in KDP. The coherence length ℓ'_{coh} defined by Eq. (5-4) is about 20 cm long. With this beam and a crystal with thickness $z = 1.23$ cm perfect phase matching for the beam is possible and the intensity of the second-harmonic generation should vary as z^2. This was verified by Ashkin, Boyd, and Dziedzic in crystals of three different thicknesses. The behavior of the second-harmonic intensity around the exact phase-matched direction is well represented by Eq. (4-13), as shown in Figure 5-4. Since the power of the cw gas laser beam is only 1.48×10^{-3} watts, the second-harmonic power is very small, in spite of the perfect

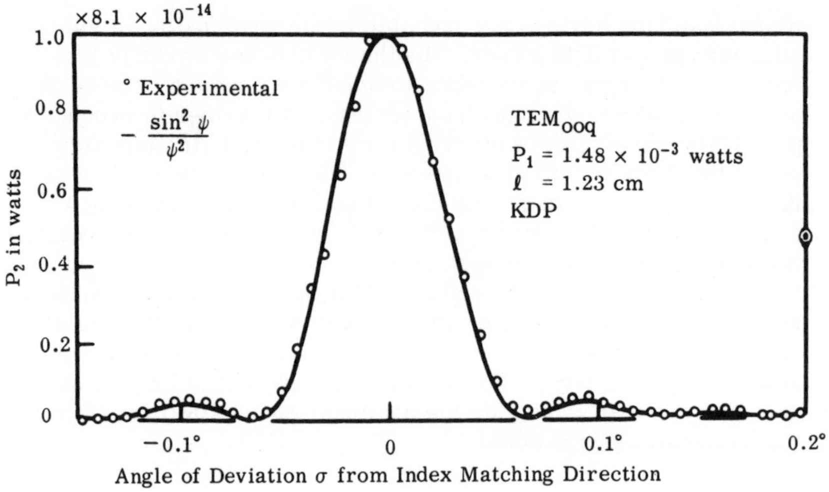

Figure 5-4. Second-harmonic generation in KDP by a gas laser in a single mode near the exact phase matched direction (after Ashkin et al.[13])

phase matching. The observed maximum second-harmonic power generated in a KDP crystal, 1.23 cm thick, is 8.1×10^{-14} watts. The polarization of the incident laser beam is along [110]. The extraordinary second-harmonic ray is polarized in the plane containing the crystallographic z-axis and the direction of phase propagation which makes an angle $\theta_M = 41°$ with the z-axis for phase velocity matching at $\lambda = 1.15$ microns. The direction of energy propagation makes a small angle α with the direction of phase propagation. If the beam diameter is a, the power in the extraordinary second-harmonic ray, with a phase velocity parallel to the fundamental ordinary ray, would run sideways out of the cross section after a distance $d/\mathrm{tg}\alpha$. With a = 0.21 cm and z = 1.25 cm this effect is clearly not important in this experiment. The effective nonlinear source term driving the extraordinary harmonic wave is the component of polarization parallel to the electric field vector $\mathbf{E}(2\omega)$ of the wave,

$$P^{NLS}(r) = P_z^{NLS} \sin(\theta_M + \alpha) = \chi_{36}(2\omega = \omega + \omega) \times \sin(\theta_M + \alpha) E_{1x}(r) E_{1y}(r)$$

This value for $P^{NLS}(r)$ is substituted into Eq. (4-13) and an integration over the precisely known radial contribution of the fundamental intensity is performed. In this manner Askhin, Boyd, and Dziedzic found $\chi_{36}^{KDP}(2\omega = \omega + \omega) = (6 \pm 2) \times 10^{-9}$ esu for $2\pi c\omega^{-1} =$ 1.15 microns. This value is twice as large as quoted in the original paper, because of our definition of the field amplitudes. SHG was also measured for other modes. The observations agreed well with the calculated transverse intensity distributions.

5-3 MULTIMODE STRUCTURE AND FLUCTUATION PHENOMENA

Although the gas laser can be operated in a single transverse mode, there may still be several longitudinal modes with equally spaced frequencies. It will be shown that this may cause a systematic error and that the true value of χ_{36} may be smaller by a factor $2^{1/2}$. The effects of the multimode structure of actual lasers, which may be considered as a number of coherent oscillators in parallel, are troublesome and cause fluctuations in the second-harmonic production. Other nonlinear phenomena should also be described in a stochastic sense, because the incident laser field is composed of a number of waves, whose amplitudes and phases are random variables. They are not necessarily statistically independent. The nonlinear processes will couple different modes and may establish partial or complete correlation between them. In some cases it is valid to consider the complex amplitudes as statistically independent. A ruby laser where different modes make predominantly use of different excited ions is an example.

Describe the field of a laser for one direction of polarization by

$$E_1(\mathbf{r},t) = \sum_{n,j} a_{nj}\, u_j(x,y) \exp(ik_{nj} z - i\omega_{nj} t) \qquad (5\text{-}11)$$

One index j has been used to label the transverse modes. The second harmonic intensity created by this laser field in a nonlinear slab of thickness z is given by Eq. (4-13) and can be written as a quadruple sum

$$|E_2(x,y,z,t)|^2 = \frac{64\pi^2 |\chi^{NL}|^2 \sin^2 \frac{1}{2}\omega c^{-1}(\epsilon_S^{1/2} - \epsilon_T^{1/2})z}{(\epsilon_S - \epsilon_T)^2}$$

$$\times \sum e^{-i(\omega_{nj} + \omega_{n'j'} - \omega_{n''j''} - \omega_{n'''j'''})t} \qquad (5\text{-}12)$$

$$\times\, a_{nj}\, a_{n'j'}\, a^*_{n''j''} a^*_{n'''j'''}\, u_j(x,y) u_{j'}(x,y) u^*_{j''}(x,y) u^*_{j'''}(x,y)$$

The experimentally observed quantity is usually detected by a quadratic process, integrated over a certain time interval and a certain cross-sectional area. If, for example, the current in a photomultiplier tube is measured, caused by the light at the second-harmonic frequency, one integrates over a time interval T, the time constant of the detecting circuit, and over the cross section of the photocathode. Although T will be short enough, that the amplitudes can be considered time independent, T will usually be longer than the period of any beat between different frequency components. The exponential function in Eq. (5-12) integrated over T may be replaced by a δ-function. Since the different modes u_j are orthonormal, the fundamental intensity or photocurrent is simply,

$$I_1 = \eta_1 \sum_{nj,n''j''} a_{nj} a^*_{n''j''} \frac{1}{T} \int_0^T e^{-i(\omega_{nj} - \omega^*_{n''j''})t}\, dt \times \iint u_{nj} u^*_{n''j''} dxdy = \eta_1 \sum |a_{nj}|^2 \tag{5-13}$$

The factor η_1 absorbs the photosensitivity and other circuit constants.

The measured second-harmonic intensity or photocurrent can in general be obtained from Eq. (5-12). In the simple case that the amplitudes of the individual modes of oscillation are constant, but the phases ϕ_{nj} are statistical independent variables uniformly distributed over the interval 0-2π, the average second-harmonic intensity can be written as

$$\bar{I}_2 = \eta_2 |\chi^{NL}|^2 \sum |a_{nj}|^4 C_{jjjj} + 2\eta_2 |\chi^{NL}|^2 \times \sum{}' |a_{nj}|^2 |a_{n'j'}|^2 C_{jj'jj'} \tag{5-14}$$

Here the constant η_2 includes the photosenstivity at the harmonic frequency and the geometrical factor in Eq. (5-12). The constants C are defined by

$$C_{ijk\ell} = \iint u_i u_j u^*_k u^*_\ell \, dxdy \tag{5-15}$$

If the amplitudes are constant, the fundamental intensity is not a random function, but the second-harmonic intensity is, because different fundamental mode pairs with random phases contribute to the same second-harmonic mode.

Consider the gas laser with many equally spaced longitudinal modes, but only one transverse mode j. The amplitudes of these modes may be considered constant, but the phases are not necessarily statistically independent. If this is assumed, combination of Eqs. (5-13) and (5-14) yields for N modes of equal amplitude,

$$\bar{I}_2 = (\eta_2/\eta_1^2)\,|\chi^{NL}|^2\, I_1^2 \iint |u_j(r)|^4\, dxdy \times \left\{ 1 + \frac{\Sigma'_{n' \neq n} |a_n|^2 |a_{n'}|^2}{(\Sigma_n |a_n|^2)^2} \right\} \tag{5-16}$$

The primed summation is over N(N − 1) pairs of modes. If a given fundamental intensity I_1 is distributed over N modes instead of in one mode, the average second-harmonic intensity is larger by a factor $(2 + N^{-1})$. The nonlinear susceptibility derived from such an experiment, without taking the mode structure into account, could be a factor $2^{1/2}$ too large.

A nonlinear effect always weights the samples with constructive interference more heavily than the samples with negative interference. Inhomogeneous distributions in time and space, for a given integrated fundamental intensity, will yield more harmonic power. Occasional deviations from the quadratic relationship $I_2 \propto I_1^2$ have been reported. The explanation is probable, that the mode structure changes as the laser power I_1 is increased.

In addition to the increased average harmonic power, this power will show statistical fluctuations around the mean. $\langle I_2^2 \rangle - \langle I_2 \rangle^2$ may be calculated in terms of an eighth power polynomial in the random amplitudes of the fundamental modes. Similar considerations hold a fortiori for higher order nonlinear processes.

Ducuing[14] has observed the stochastic relationship between I_2 and I_1^2 for individual pulses of a Q-switched ruby laser. Sometimes a pulse with lower fundamental intensity gives more harmonic yield. There is, however, a perfect nonrandom one-to-one correspondence between two nonlinear processes of the same order, obtained from the same laser pulse by means of a partially reflecting mirror. In this manner reliable *relative* values of the nonlinear susceptibility can be obtained. The experimental arrangement is shown in Figure 5-1, where the nonlinearity of GaAs is measured relative to the SHG in KDP or quartz. If a nonabsorbing ground-glass scatterer is inserted in front of the quartz crystal, the one-to-one correspondence is lost. The spatial distribution of modes in the two nonlinear samples is not any more identical. This experiment shows conclusively that in a ruby laser pulse several spatial modes go at the same time.

The multimode structure also has a profound effect on the generation of small beat frequencies. Consider first the influence of temporal coherence in two beams with N_1 and N_2 frequency components, respectively, centered around ω_1 and ω_2. Each component of the first beam will beat with each component of the second beam to give a difference near $\omega_2 - \omega_1$. The bandwidth of the detection system at the microwave frequency $\omega_2 - \omega_1$ will determine how many of these $N_1 N_2$ beats will contribute to the signal. If the bandwidth of the detector is smaller than the equal spacing between the individual N components in each light beam, the resultant signal at $\omega_2 - \omega_1$ will be reduced by a factor equal to the larger number, N_1 or N_2, compared to the signal produced by two purely monochromatic light beams of the same intensity.

The spatial coherence effect is similar. Suppose that the wave length at the difference frequency is equal to or longer than the cross section of the beams or the nonlinear crystal. The power generated at the difference frequency is then proportional to the square of the average polarization at $\omega_2 - \omega_1$ over the cross section. A spatial distribution of polarization with nodes in a cross section with dimensions small compared to the wave length, cannot radiate. There is destructive interference. Therefore, only modes in the two light beams with the same spatial distribution can produce intensity at a difference beat frequency, when the wave length is longer than the diameter of the light beams. If there are J_1 and J_2 spatial modes in the two beams, the difference beat intensity is reduced by a factor equal to the larger number, J_1 or J_2. In creating microwave beats from light, it is important to have diffraction limited beams. The reduction factor is essentially due to limitations of final states in phase space at low frequencies.

In the inverse process of microwave modulation of an incoherent light beam the reduction factor does not apply. The light components in the side bands can always find a mode with the appropriate frequency $\omega_{nj} \pm \omega_{micro}$ and an appropriate cross-sectional distribution $u_{nj} u_{micro}^{(*)}$ of the amplitude. The reduction factor clearly does not apply in the mixing experiments of ruby laser light with neodymium laser light with a wide frequency distribution or mercury incoherent light.

The reduction factor does not apply either for dc rectification of light. In that case each mode can beat with itself to give a dc voltage proportional to the integrated fundamental intensity, regardless of the mode distribution.

The situation is essentially identical to the discussion of coherence effects in photoelectric mixing given by Forrester.[15] The photoelectric effect is also a nonlinear effect with a quadratic response.

5-4 NONLINEAR SUSCEPTIBILITIES OF PIEZOELECTRIC CRYSTALS

The methods outlined in Chapters 1 and 5 have been used to measure the third-rank susceptibility tensors describing SHG in a number of crystals. The data are usually obtained relative to the

TABLE 5-1

Relative Values of the Nonlinear Susceptibility Elements $\chi_{ijk}(2\omega = \omega + \omega)$. The Voigt Notation is used, x = 1, y = 2, z = 3, yz = zy = 4, xz = zx = 5, xy = yx = 6. The Reference Value is $d_{36} = 12 \times 10^{-9}$ esu for KDP.[13]

Crystal	Susceptibility for SHG		Reference
	λ = 6943 Å	λ = 1.06 microns	
KH_2PO_4	χ_{36} = 1.00	1.0	
(tetragonal)	χ_{14} = 0.95 ± 0.06	1.01 ± 0.05	5 – 16
	χ_{14} = 0.86 ± 0.05		5 – 17
KH_2PO_4	χ'_{31} = 1.23 ± 0.1		5 – 17
(orthorhombic)	χ'_{15} = 1.20 ± 0.1		5 – 17
	χ'_{32} = 0.83 ± 0.1		5 – 17
	χ'_{24} = 0.80 ± 0.1		5 – 17
	χ'_{33} = 0.0		5 – 17
KD_2PO_4	χ_{36} = 0.75 ± 0.02	0.92 ± 0.04	5 – 16
	χ_{14} = 0.76 ± 0.04	0.91 ± 0.03	5 – 16
	χ_{14} = 0.85 ± 0.1		5 – 17
$NH_4H_2PO_4$	χ_{36} = 0.93 ± 0.06	0.99 ± 0.05	5 – 16
	χ_{14} = 0.89 ± 0.04	0.98 ± 0.05	5 – 16
	χ_{36} = 1.25 ± 0.05		5 – 17
	χ_{14} = 1.07 ± 0.05		5 – 17
KH_2AsO_4	χ_{36} = 1.0 ± 0.1		5 – 17
	χ_{14} = 0.86 ± 0.1		5 – 17
Quartz	χ_{11} =	0.82 ± 0.04	5 – 18
	χ_{14} =	0.00 ± 0.05	5 – 18
$BaTiO_3$	χ_{15} =	35.0 ± 3	5 – 18
	χ_{31} =	37.0 ± 3	5 – 18
	χ_{33} =	14.0 ± 1	5 – 18
CdS	χ_{15} =	35.0 ± 3	5 – 19
	χ_{31} =	32.0 ± 2	5 – 19
	χ_{33} =	63.0 ± 4	5 – 19
GaAs	χ_{14} = 570 ± 100	330.0 ± 100	5 – 20
InAs	= 370 ± 75	210.0 ± 75	5 – 20
InSb	= 820 ± 150	324.0 ± 100	5 – 20
GaSb	= 865 ± 150	224.0 ± 75	5 – 20
ZnO	$\chi_{24} = \chi_{15}$	4.7 ± 0.4	5 – 21
	$\chi_{32} = \chi_{31}$	4.3 ± 0.4	
	χ_{33} =	14.3 ± 0.4	
$AlPO_4$	χ_{11} =	0.84 ± 0.07	5 – 21

standard value for KDP at room temperature $\chi_{zxy}(2\omega = \omega + \omega) = 6.0 \times 10^{-9}$ esu = 1.00 on the relative scale. In obtaining accurate relative values, it is important to avoid phase-matched or nearly phase-matched directions. There the data would be extremely sensitive to the detailed geometry, and very precise knowledge of the mode configuration and crystal orientation would be necessary. Reliable data in transmission furthermore require the establishment of several maxima and minima in curves of the type shown in Figure 1-2. Then Eq. (4-13) can be used for various orientations of the polarizations of the fundamental and second-harmonic beams with respect to the crystallographic axes, to determine all independent elements of the susceptibility tensor. Data have been obtained almost exclusively with ruby lasers with $\hbar\omega$ = 1.79 ev (λ = 6940 Å), or Nd^{3+} lasers in glass or $CaWO_4$, with $\hbar\omega$ = 1.17 ev (λ = 1.06 micron). Only the more recent and probably more reliable data are listed in Table 5-1.

Since two frequencies are equal, the symmetry relation exists $\chi_{iyz}(2\omega = \omega + \omega) = \chi_{izy}(2\omega = \omega + \omega) = \frac{1}{2}d_{i4}$. Here i runs over x = 1, y = 2, and z = 3. The Voigt notation has been used with the symbolism yz = zy = 4, xz = zx = 5, xy = yx = 6. The first four entries have the tetragonal symmetry, $\bar{4}2$ m or V_d, of KDP. The only nonvanishing tensor elements are $d_{14} = d_{25}$ and d_{36}. Trigonal quartz has the symmetry 32 or P_3 with nonvanishing elements $d_{11} = -d_{22} = -d_{26}$ and $d_{14} = -d_{25}$. The last four entries are III-V semiconductors with the tetrahedral ZnS symmetry, $\bar{4}3$ m or T_d. The only nonvanishing elements are $d_{14} = d_{25} = d_{36}$. For the symmetry properties of the other crystals the references given in Table 1 should be consulted. The third rank tensor d has the same symmetry properties as the piezoelectric tensor. It is often used in the literature instead of χ. Its elements in 4, 5, 6, are twice as large.

The nonlinear optical experiments are in agreement with the known symmetry properties. It is interesting to note that an additional symmetry property, known as the Kleinman relationship,[22] is often valid. The frequency dispersion of a nonlinear susceptibility in the region where the crystals are transparent is so small that 2ω and ω may also often be interchanged *without* the concomitant change of indices. This provides the additional relationship $d_{14} \approx d_{36}$ in the KDP class and $d_{14} \approx d_{25}$ in quartz. Since crystal symmetry requires $d_{14} = -d_{25}$, one finds that $d_{14} = 0$ in quartz, within the experimental limit of accuracy. The relationship in KDP is probably not satisfied rigorously as van der Ziel finds $d_{14} = (0.86 \pm 0.05)\, d_{25}$.

Temperature Dependence and Dispersion of the Nonlinear Susceptility in KDP

Van der Ziel[17] has made a thorough investigation of the temperature dependence of the nonlinear susceptibility in KDP and some crystals with the same structure. KDP undergoes a transition to a ferroelectric phase with orthorhombic symmetry below 122°K. When the crystal is cooled through the transition temperature with a sufficiently strong electric field applied along the tetragonal axis, a single ferroelectric domain results with the orthorhombic axes, x′, y′, turned 45° with respect to the original x, y axes in the tetragonal phase. The nonlinear tensor in the symmetry mm 2 of the ferroelectric phase is,

	x′x′	y′y′	zz	zx′	y′z	x′y′
x′	0	0	0	0	χ'_{15}	0
y′	0	0	0	χ'_{14}	0	0
z	χ'_{31}	χ'_{32}	χ'_{33}	0	0	0

The measured susceptibilities are shown in Figure 5-5. They are independent of temperature in each phase, χ'_{33} in the orthorhombic phase was too small to be measurable. The elements χ'_{31} and χ'_{15} have probably opposite sign, so do χ'_{32} and χ'_{24}. The absolute change in the susceptibility at the transition temperature is small. In the tetragonal phase with a fourfold axis of rotation inversion, these same quantities referred to the orthorhombic axes are required to have opposite sign by symmetry.

When the crystal is cooled without an applied electric field, a ferroelectric domain structure appears with the orthorhombic axes in adjacent domains turned by 90°. This reverses the sign of the nonlinear susceptibility. Since the domain size is comparable to the coherence length, it is possible to get strongly enhanced yields in the multidomain crystal. Just as the phase difference between the harmonic and fundamental wave would get too large, the relative sign of the induced second-harmonic polarization with respect to the waves changes. The harmonic wave continues to be enhanced. The SHG in multidomain crystals may be enhanced by up to two orders of magnitude and depends on the detailed domain structure. Similar effects have been observed by Miller[18] in ferroelectric $BaTiO_3$. The same effect may be used to achieve quasi-phase

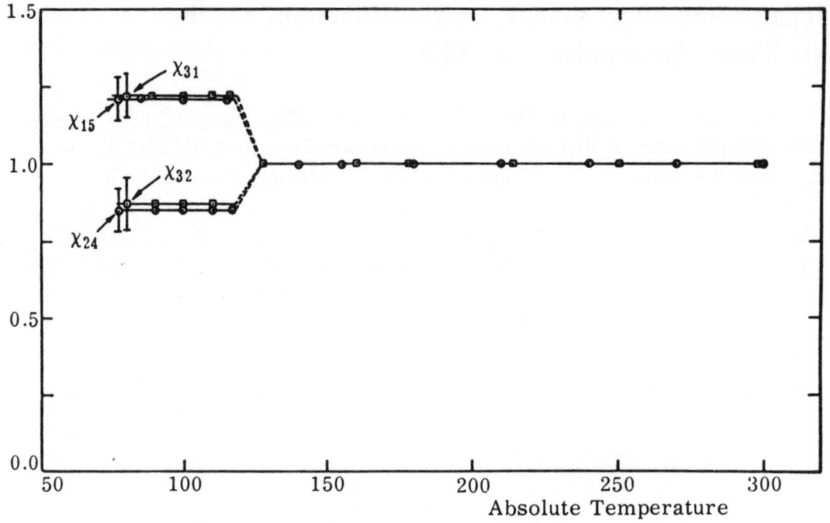

Figure 5-5. Temperature dependence of the nonlinear susceptibility in KDP. A ferroelectric phase transition occurs at 122° K.

matching in cubic crystals by stacking platelets with the 011 axes alternately turned by 90°, as mentioned in Appendix 1.

The nonlinear constant $\chi_{36}(0 = \omega - \omega)$ is about 40 times larger than $\chi_{36}(2\omega = \omega + \omega)$. This large dispersion is caused mostly by the enhancement of local fields by the ionic polarization and has been discussed extensively by Franken.[10] At microwave frequencies this ionic enhancement is still present $\chi_{36}(\omega_{\text{micro}} = \omega_2 - \omega_1) = \chi_{36}(0 = \omega_1 - \omega_1)$.

It is possible to relate the nonlinear susceptibilities in the orthorhombic phase to a higher order nonlinear susceptibility in the tetragonal phase with an internal dc field added caused by the spontaneous static polarization. One has, for example,

$$\chi'_{113}(2\omega = \omega + \omega) = \chi_{123}^{\text{tet}}(2\omega = \omega + \omega) + \chi_{1233}^{\text{tet}}(2\omega = \omega + \omega + 0)\, P_3 + \chi_{12333}^{\text{tet}}(2\omega = \omega + \omega + 0 + 0)\, P_3^2$$

$$\chi'_{223}(2\omega = \omega + \omega) = -\{\chi_{123}^{\text{tet}}(2\omega = \omega + \omega) - \chi_{1233}^{\text{tet}}(2\omega = \omega + \omega + 0)\, P_3 + \chi_{12333}^{\text{tet}}(2\omega = \omega + \omega + 0 + 0)\, P_3^2\}$$

In a similar way the birefringence in the orthorhombic phase has been related to the indices of refraction in the tetragonal phase with a linear electro-optic Kerr effect superimposed, due to the local electric field from the spontaneous static polarization. Van der Ziel[17] and Miller[18] have drawn some interesting conclusions about the local field from the relationships between these tensors.

Nonlinear Susceptibilities of Semiconductors

The dispersion of the nonlinear susceptibility is more pronounced when one of the frequencies enters into the absorption region of the material. If the second-harmonic frequency becomes absorbed, a marked increase in the nonlinear susceptibility occurs. This effect was observed by Soref and Moos,[19] who worked with ZnS—CdS and CdS—CdSe alloys. The band gap E_g in these crystals varies in a systematic manner with the concentration, from 1.71 ev in CdSe to 2.36 ev in CdS, and to 3.52 ev in ZnS. This range encompasses twice the photon energy of a Nd-glass laser. The nonlinear susceptibility $\chi_{333}(2\omega = \omega + \omega)$ increases by an order of magnitude as the band gap E_g is decreased from $1.52(2\hbar\omega)$ to $0.73(2\hbar\omega)$, thus entering the region of strong second-harmonic absorption. This result is consistent with the theoretical expression Eq. (2-48). The resonance denominator at $\omega_1 + \omega_2 = 2\omega$ in this equation shows that the nonlinear susceptibility should increase in a similar manner as the complex linear susceptibility at 2ω, when $2\hbar\omega$ becomes larger than E_g. There is at least a qualitative agreement between theory and experiment. It is noteworthy that even in the transparent region $E_g > 2\hbar\omega$ the nonlinearity of CdS is much larger than KDP. A probable explanation is that the deviation from inversion symmetry is much greater in CdS and the III-V semiconductors. Therefore, considerable more oscillator strength is associated with wave functions of valence electrons, which lack a well-defined parity.

Although all these nonlinearities have been measured in transmission, the nonlinear susceptibility of the III-V semiconductors has been measured with the reflection technique shown in Figure 5-1. Since the fundamental laser frequency is absorbed by the crystal, the incident power flux must not be too high in order to avoid damage of the surface by excessive heating and evaporation. Fortunately a large fraction of the absorbed laser light is probably re-emitted as recombination radiation. The surface should be polished and chemically etched. All reflected harmonic radiation originates in a layer about 400 Å thick, the absorption depth at the second-harmonic frequency. A test for the quality of this layer is the zero of the harmonic intensity for the orientations shown in Figures 5-2 and 5-3. This proves that the layer has the true symmetry of the bulk crystal. With the laser beam incident at 45° and a polarization normal to the plane of incidence, parallel to the crystallographic [111] axis, the nonlinear polarization is,

$$P_{\perp}^{NLS}(2\omega) = 3^{-1/2}\chi_{14}|E_{\perp}(\omega)|^2 = 3^{-1/2}|E^{inc}(\omega)|^2 \times \left(\frac{2}{(2\epsilon(\omega)-1)^{1/2}+1}\right)^2 \qquad (5\text{-}17)$$

The last factor is the Fresnel transmission coefficient used to express the field inside the semiconductor in terms of the field strength of incident laser beam. With the use of Eq. (4-8), the reflected harmonic amplitude is

$$E_{\perp}^{R}(2\omega) = (4\pi\chi_{14}/3^{1/2})|E^{inc}(\omega)|^2 \left(\frac{2}{(2\epsilon(\omega)-1)^{1/2}+1}\right)^2 \times \frac{2^{1/2}}{\{(\epsilon(2\omega)-\frac{1}{2})^{1/2}+(\epsilon(\omega)-\frac{1}{2})^{1/2}\}\{(2\epsilon(2\omega)-1)^{1/2}+1\}} \qquad (5\text{-}18)$$

The observed second-harmonic intensity is proportional to the squared absolute value of this expression. It should be emphasized that only the absolute value of the complex susceptibility is measured in SHG experiments.

The band gap in GaAs is 1.35 ev, larger than the Nd^{3+} laser frequency at 1.17 ev. In this case the nonlinear susceptibility can also be measured in transmission with Eq. (4-15). The two methods agree, but the reflection method is two to three times more accurate. A large uncertainty in the measurement is caused by the uncertainty in the complex linear index of refraction constants. In the transmission formula (Eq.4-15) a difference of this quantity at two frequencies ω and 2ω is involved. This limits the accuracy of the transmission method.

There is considerable dispersion in $\chi_{14}(2\omega = \omega + \omega)$ between $\hbar\omega = 1.17$ ev and 1.79 ev. For InSb with $E_g = 0.24$ ev, InAs with $E_g = 0.45$ ev, and GaSb with $E_g = 0.81$ ev, both fundamental and second harmonic are always absorbed. In GaAs with $E_g = 1.35$ ev the fundamental is not absorbed at the Nd laser frequency. It is noteworthy that nothing spectacular happens to the nonlinear susceptibility when the fundamental becomes absorbed in addition to the second harmonic. There is a big increase by an order of magnitude when that frequency passes into the absorption region, but additional absorption at the fundamental at best changes the slope of the dispersion, not the order of magnitude of the susceptibility. This behavior is again in general agreement with Eq. (2-48). Since the bands are broad in comparison to the damping constant Γ, one may take the limit $\Gamma \to 0$ in the integral and use the relation,

$$\lim_{\Gamma \to 0} \frac{1}{\omega_2 - \omega_{n''n'} + i\Gamma} = \text{Principal Value} \frac{1}{\omega_2 - \omega_{n''n'}} - i\pi\, \delta(\omega_2 - \omega_{n''n'}) \tag{5-19}$$

to evaluate the integral in Eq. (2-48). In this manner the nonlinear susceptibilities can be expressed in the same joint densities of state that occur in the theory of the linear dielectric constant.[23] There will be critical points, where this joint density of states is high. When the nonlinearities can be measured at higher frequencies, such maxima in the complex nonlinear susceptibility may be looked for. Inspection of Eq. (2-48) shows that another kind of singular behavior is possible that does not occur in the linear case. Two denominators may be near resonance at the same time, when $\omega_{n'n}(k) = 2\omega_{n'n''}(k)$. In this case, and also near the band edge, Eq. (2-48) should be used and the limit $\Gamma \to 0$ of Eq. (5-19) should not be taken. Although the nonlinearities in these semiconducting compounds are very large, the actual SHG is small. The effective coherent volume of radiating polarization, proportional to the absorption depth, decreases faster than the nonlinearity increases. The nonlinear susceptibility gives more information about the electronic band structure than can be obtained from the linear optical constants alone, because it measures properties that are peculiar to the mixed parity of bandwave functions.

The semiconductors Si and Ge that possess a center of inversion have no measurable SHG. Their nonlinearity is at least three orders of magnitude smaller than in GaAs and the second-harmonic intensity less by a factor 10^6. The effects are independent of the doping and resistivity of the materials, indicating that nonlinearities of the conduction electron plasma are unobservable. This is not surprising. The nonlinearity of the plasma in a metal such as Ag is estimated according to Eq. (1-6) or Eq. (2-45) to be of the same order as KDP, or about four hundred times smaller than in GaAs. Even this will be difficult to observe. The absorption depth is very small in a metal. Even if one uses the sharp plasma resonance in Ag to make some of the denominators that occur in expressions similar to Eqs. (5-17) or (5-18) as small as possible, it will still be a difficult experiment. The nonlinearity of a semiconductor plasma can be enhanced by a large factor in the far infrared for small effective masses and in high magnetic fields, when cyclotron resonance effects occur.[24] There is clearly much opportunity for further experimentation.

5-5 ELECTRIC QUADRUPOLE EFFECTS

Terhune[25] has observed SHG in a calcite crystal, which has a center of inversion. The point group symmetry is $\bar{3}m$ or D_{3d}. The effect is quite small and is observed only by careful momentum matching of the fundamental ordinary wave and the extraordinary wave at 2ω. The SHG can be enhanced greatly by an applied dc electric field, as shown in Figure 5-6. This field destroys the inversion symmetry. One can express the source term in this case by a fourth-rank tensor,

$$\mathbf{P}^{NLS}(2\omega) = \chi(2\omega = \omega + \omega + 0)\,\mathbf{E}(\omega)\,\mathbf{E}(\omega)\,\mathbf{E}_{dc}(0) \qquad (5\text{-}20)$$

The second-harmonic intensity indeed increases as E_{dc}^2. In this section the interest is centered on the case $E_{dc}(0) = 0$. The SHG must be caused by a quadrupole effect derivable from an energy function,

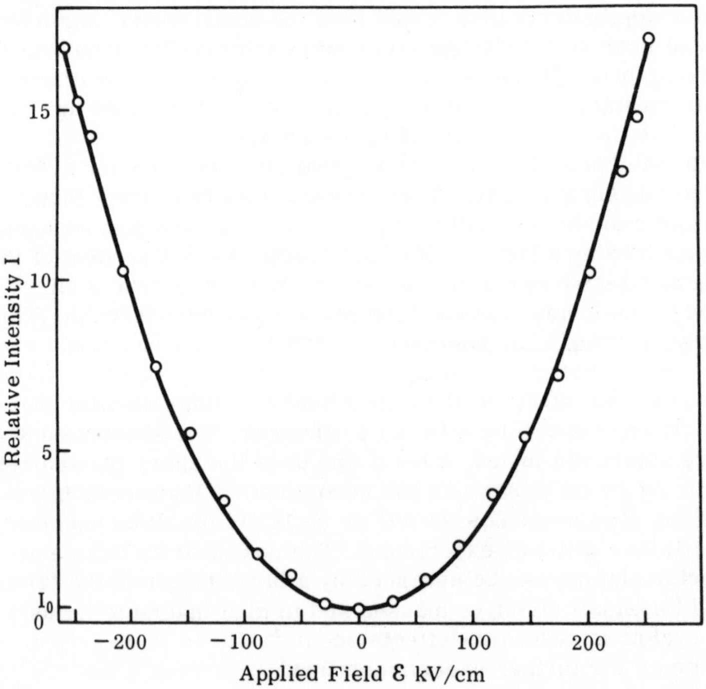

Figure 5-6. Second-harmonic generation in calcite as a function of an applied dc electric field. Without applied field, only a quadrupole effect remains (after Terhune).[39]

$$F = -\sum_{ijk\ell} \chi^{I}_{ijk\ell} E_i^*(2\omega)E_j(\omega)\nabla_k E_\ell(\omega) - \sum_{ijk\ell} \chi^{II}_{ijk\ell} E_j(\omega)E_\ell(\omega)\nabla_k E_i^*(2\omega) + \text{complex conj.} \tag{5-21}$$

The susceptibility tensor is symmetric in the last two indices. Pershan[26] has discussed the quadrupole energy terms in detail, and he has shown that one may put $\chi^{I}_{ijk\ell} = 0$ without loss of generality. In this case the nonlinear source current density is, according to Eq. (3-3),

$$J_i(2\omega) = 2i\omega \sum_{\ell jk} \nabla_k \chi^{II}_{j\ell ki} E_j(\omega)E_\ell(\omega) \tag{5-22}$$

For plane waves the gradient ∇_k is only nonvanishing in the direction of the wave and since $i = k$, the phase-matched second-harmonic wave has only a longitudinal source term. In an isotropic medium this term could not generate a wave, but in calcite the angle α between **D** and **E** for the phase-matched direction is about 7°. The longitudinal current can thus drive the wave with an effective component,

$$J_{\parallel}^{NLS}(2\omega) = J_{i=k}(2\omega) \sin \alpha$$

There are several tensor components in Eq. (5-22) that contribute to the source term in the geometry of the experiment. The order of magnitude of these elements may be estimated from the intensity observed by Terhune, $|\chi^{II}| \sin \alpha \approx 10^{-18}$ esu or $|\chi^{II}| \sim 10^{-17}$ esu. This is clearly only an order of magnitude estimate, since the difficulties in obtaining reliable absolute values are even more severe than in the preceding cases.

In isotropic media SHG can take place by the quadrupole interaction, if two light beams traveling at an angle φ to each other and with different directions of polarization cross in the interaction volume. For momentum matching it is still desirable to use an anisotropic crystal with the second harmonic propagating at angles φ_1 and $\varphi_2 = \varphi - \varphi_1$ with respect to the two fundamental beams. The nonlinear source Eq. (5-22) has transverse components in this case. Giordmaine[27] has described an experiment in calcite using this geometry.

Magnetic dipole effects are also derivable from a suitable energy function, one example of which occurs in Eq. (3-11). Pershan[26] has given an extensive discussion of its symmetry properties. It should be possible to observe an induced magnetization by a circularly polarized light pulse in a nonabsorbing medium. This effect is the thermodynamic inverse of the Faraday rotation and is derivable from the same energy with Verdet's constant. This effect has not yet been reported.

A large magnetization can be induced by a circularly polarized ruby laser beam, absorbed in another ruby. The populations of the Zeeman states of the Cr^{3+} ions are redistributed in this case, which is closely related to optical polarization pumping in gases.

5-6 THIRD-HARMONIC GENERATION

In the remainder of this chapter effects will be considered that are derivable from a time-averaged energy which is a fourth-degree polynomial in the electric field strengths. Terhune and coworkers[28] have observed third-harmonic generation, THG, in isotropic liquids, cubic crystals, and in calcite. In the last case momentum matching of the fundamental beam and third-harmonic beam is possible in two directions. At 57° from the trigonal axis, one has $2k_o(\omega) + k_e(\omega) = k_e(3\omega)$, and at 47° the relation $3k_o(\omega) = k_3(3\omega)$ holds. Here the subscripts o and e refer to the ordinary and extraordinary modes of polarization. The THG is described by a fourth-rank tensor,

$$P_i(3\omega) = \sum_{jk\ell} \chi_{ijk\ell}(3\omega = \omega + \omega + \omega)E_j(\omega)E_k(\omega)E_\ell(\omega) \qquad (5\text{-}23)$$

Terhune[28] has given the form of this tensor in the 32-point group symmetries. The observed intensity at 3ω is indeed found to be proportional to the cube of the fundamental intensity, as shown in Figure 5-7. The maximum energy conversion efficiency is three parts per million. This is obtained in a focused beam from a Q-switched laser along a phase-matched direction. The intensity of the field at the focus is so large that almost any crystal can be damaged. The intensity of the laser beam is reduced just below the critical value to observe THG. Damage may occur due to absorption at impurities, multiple photon absorption, or the generation of vibrational energy by Raman or Brillouin scattering processes. The sudden heat production in a small volume produces a shock wave. The estimated magnitude of $|\chi(3\omega = \omega + \omega + \omega)|$ is 10^{-15} esu.

In cubic crystals such as LiF, Eq. (5-23) can be written in the form

$$P_i = AE_i^3 + BE_i(\mathbf{E}\cdot\mathbf{E}) \qquad (5\text{-}24)$$

with $A + B = \chi_{1111}$ and $B = \chi_{1122} = \chi_{1133}$. Terhune observed that

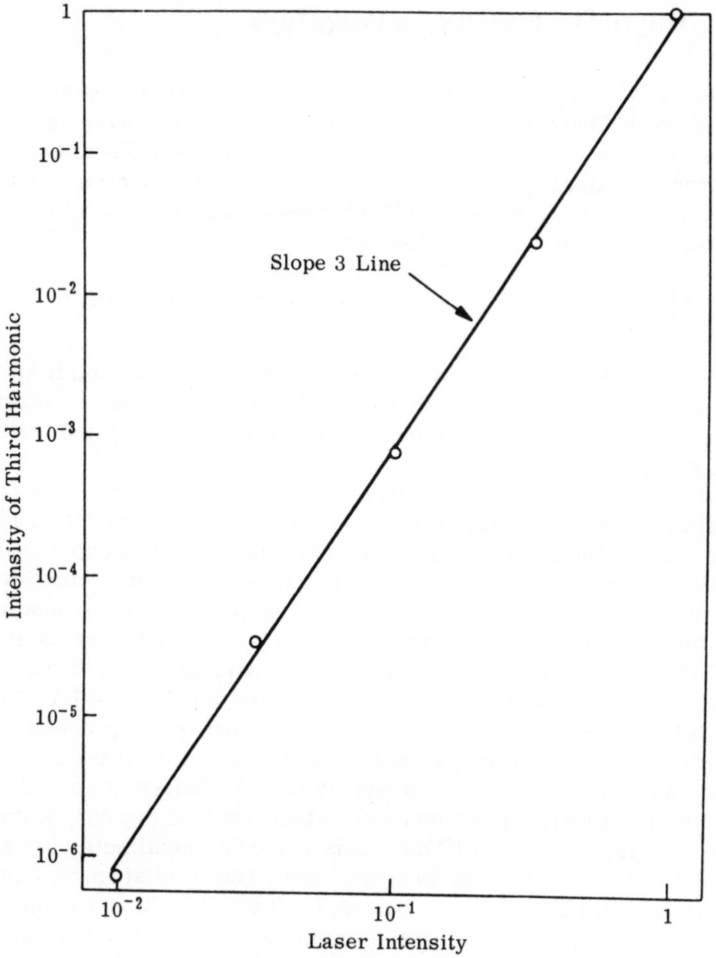

Figure 5-7. Third-harmonic generation in calcite as a function of the fundamental input power (after Terhune et al.).[28]

THG shows some anisotropy in these crystals. Since momentum matching is not possible, the effect is difficult to observe. The constant A is definitely different from zero.

5-7 MULTIPLE PHOTON ABSORPTION

The cross section for two photon absorption processes has been discussed in Chapter 2. Kleinman[29] and Braunstein[30] have analyzed the problem in more detail. If the incident light beam is coherent, the process can also be described by an imaginary nonlinear susceptibility as shown in section 2-8. The two quanta absorption from the same light wave is described by

$$P_i^{NL}(\omega) = i\chi''_{ijk\ell}(\omega = -\omega + \omega + \omega)E_j^*(-\omega)E_k(\omega)E_\ell(\omega) \quad (5\text{-}25)$$

The symmetry properties of this tensor, which is symmetric in the first two and in the last two indices, are the same as the symmetry properties of the elastic tensors, which have been tabulated for all point groups.

Two photon absorption was first observed by Kaiser and Garrett[31] who illuminated a crystal of CaF_2 containing Eu^{2+} ions with a ruby laser beam. The crystal emitted a characteristic fluorescence of an excited state of Eu^{2+} at 4250 Å. The intensity of the fluorescence was proportional to the square of the intensity of the laser beam. This blue fluorescence was previously only observed after excitation in the UV absorption band of Eu^{++}. Apparently two red quanta are able to induce a transition to this excited level in the UV. The possibility of two successive single photon absorption process is excluded. There is no measurable linear absorption at the ruby frequency and there are no known energy levels "half-way up." About 1 in 10^7 of the incident photons were absorbed in a crystal, 1mm thick, with $N_0 = 2.4 \times 10^{18}$ Eu^{2+} ions per cubic centimeter for a flux density of 8×10^{23} photons/cm² sec. The field strength can be calculated from Eq. (2-91) to be $|E_\lambda| = 9$ esu = 2700 volts/cm. The nonlinear susceptibility follows from the absorption per centimeter, $4\pi(\omega_\lambda/c)(+\chi'')|E_\lambda|^2 = 10^{-6}$ or $\chi''(\omega = -\omega + \omega + \omega) \approx 10^{-14}$ esu. This value could be considerably enhanced if the absorption width of the UV band could be reduced and the concentration of Eu^{2+} ions increased. The susceptibilities at a sharp resonance can be larger by a significant factor.

An important spectroscopic application of multiple photon absorption has been introduced by Hopfield and coworkers.[32] His experimental arrangment is shown in Figure 5-8. There are two light beams, one is the ruby laser pulse, the other is a continuous xenon arc light source. Double photon absorption is detected as a change in absorption coefficient of the sample, a KI crystal, for the xenon light, when the ruby laser pulse is on. In this manner the dispersive properties of two photon absorption can be investigated over a wide frequency range. The results complement the

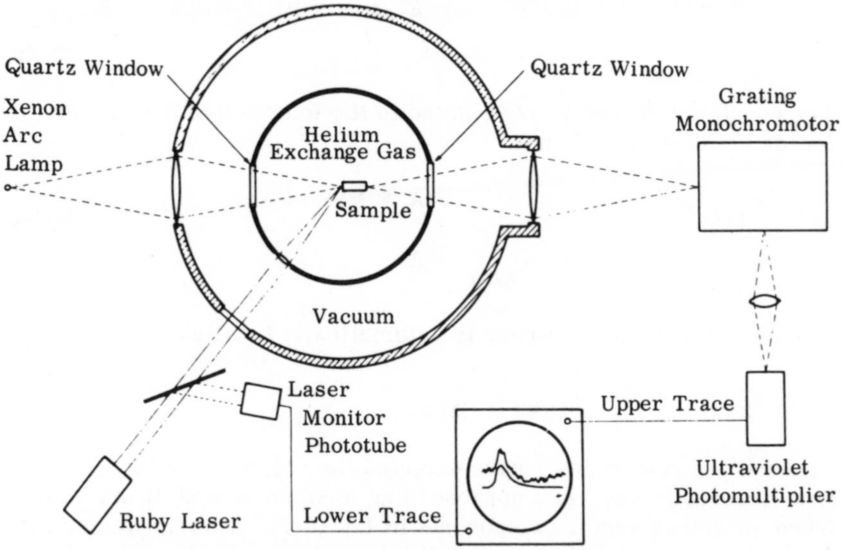

Figure 5-8. Two photon absorption spectroscopy. The change in absorption of light in a KI crystal during the laser pulse is measured. The absorption of light from the xenon arc is measured as a function of wave length (after Hopfield et al.).[22]

single photon absorption data, because in the two photon electric dipole transitions the initial and final state have the same parity.

Since the second light source is incoherent with a broad frequency spectrum, a nonlinear susceptibility, defined in Chapter 2 for periodic perturbations, is not immediately appropriate. The nonlinear absorption coefficient can of course still be expressed in terms of an absorption cross section proportional to the intensity of the incoherent beam per unit frequency interval.

Recently Singh and Bradley[33] have observed a three photon absorption in crystals of naphthalene. The fluorescent intensity at 3000 Å induced by a focused ruby laser beam was proportional to the third power of the fundamental intensity. For a photon flux density of 2×10^{27} photons/cm^2 per sec in the focus, the conversion efficiency was about $1{:}10^{10}$. A stepwise excitation process appears to be ruled out.

5-8 INTENSITY DEPENDENT INDEX OF REFRACTION

If in Eq. (5-25) the susceptibility is taken as a real quantity, one has the phenomenological description of the effect, where the index of refraction changes with intensity. In an isotropic medium the nonlinear polarization can be written in the form

$$P_i^{NL}(\omega) = AE_i(\omega)[\mathbf{E}^*(-\omega)\cdot\mathbf{E}(\omega)] + \tfrac{1}{2}BE_i^*(-\omega)[\mathbf{E}(\omega)\cdot\mathbf{E}(\omega)] \tag{5-26}$$

The constants A and B are related to the fourth-rank symmetric tensor elements as follows,

$$\begin{aligned} \chi'_{xxxx}(\omega = -\omega + \omega + \omega) &= A + \tfrac{1}{2}B \\ \chi'_{xyxy} &= A \end{aligned} \tag{5-27}$$

and the condition for isotropy is automatically fulfilled,†

$$\chi'_{xxyy} = \tfrac{1}{2}B = \chi'_{xxxx} - \chi'_{xyxy}$$

Terhune and coworkers[34] have detected the existence of the term B in an ingeneous way. In a nonabsorbing medium A and B are real. When the linear vector components in Eq. (5-26) are transformed to a circular representation $P_\pm = (P_x \pm iP_y)/\sqrt{2}$ and $E_\pm = (E_x \pm iE_y)/\sqrt{2}$, one finds for the intensity dependent change in the refractive indices for the two circular polarizations,

$$\begin{aligned} \delta n_+ &= (2\pi/n_0)[AE_+E_+^* + (A + B)E_-E_-^*] \\ \partial n_- &= (2\pi/n_0)[AE_-E_-^* + (A + B)E_+E_+^*] \end{aligned} \tag{5-28}$$

The refractive index for circular and linear polarized light is apparently different. Use elliptically polarized light from a ruby laser, focused into a liquid, through a $\frac{1}{8}\lambda$ mica plate. A Rochon prism is used to separate two linear polarizations, corresponding to the fast and slow axes of the $\frac{1}{8}\lambda$ plate. At low intensities of the laser beam the intensity in these two polarizations is equal. At high intensities an unbalance is observed due to a rotation of the polarization ellipsoid, described by Eq. (5-28). Terhune et al. have measured the nonlinear constant $B = 2(\chi'_{xxxx} - \chi'_{xyxy})$ in many liquids. The values range from 8×10^{-15} esu for water to 10^{-13} esu for carbon disulfide with our amplitude convention. The latter value is so high, because one approaches a resonance for two photon absorption. Giordmaine[35] has indeed observed the intensity dependent absorption $\chi''(\omega = -\omega + \omega + \omega)$ for ruby light in this liquid. Unfortunately, it is difficult to extend the method to cubic crystals. The strain birefringence in solids is the stumbling block.

†Sometimes the off-diagonal elements are taken half as large. In that case one writes $\chi_{xyxy} + \chi_{xyyx}$ instead of our χ'_{xyxy} in Eq. (5-27).

5-9 STIMULATED RAMAN EFFECT

Woodbury and Ng[36] observed that a ruby laser switched with a nitrobenzene Kerr cell, as shown in Figure 4a, emitted copious light at 7670 Å in addition to the normal ruby laser light at 6943 Å. The correct explanation was given shortly afterwards by Hellwarth[37] et al. When different liquids are inserted between the mirrors of the laser resonator, different frequencies are emitted. They are displaced from the ruby frequency by an amount corresponding to a vibrational frequency of each molecule.

The Raman laser spectra of several liquids[37] and solids[38] are listed in Table 5-2. Only the frequencies belonging to the sharpest

TABLE 5-2
Comparison of the Frequency Shifts of Stokes Lines Emitted in the Stimulated Raman Effect with the Frequency Shift of the Most Intense Spontaneously Emitted Raman Lines.[37,38]

Material	Shift of Coherent Raman Lines from the Ruby Frequency in cm^{-1}	Shift of Two Strongest Incoherent Stokes Lines in cm^{-1}
Benzene	3064 ± 2	3064
	990 ± 2	991.6
	1980 ± 4	
Nitrobenzene	1344 ± 2	1345 and 1004
	$2\times(1346 \pm 2)$	
	$3\times(1340 \pm 5)$	
Toluene	1004 ± 4	1002 and 785
1-Bromonaphthalene	1368	1363 and 3060
Pyridine	992 ± 2	991 and 3054
	$2\times(992 \pm 5)$	
Cyclohexane	2852 ± 1	801 and 2853
Deuterated Benzene	944.3 ± 1	944.7 and 2292
	2×944	
Diamond	1325	1331.8
	2661	
Calcite	1075	1085.6
	2171	
α-Sulphur	216	216 and 468
	472	
	946	

and most intense spontaneous Raman line show up in the stimulated process. Occasionally two lines may show up. They usually belong to a totally symmetric vibration. This could be expected as the process with the lowest threshold will tend to limit the laser power below higher threshold values. Further increase in pump power will increase the intensity of this stokes line, but not create other lines of the spontaneous Raman spectrum. When, however, the stokes line has attained a sufficient intensity, it can in turn create a stokes line of the first stokes line at $\omega_L - 2\omega_v$, and so on. These higher order lines were also reported in the first communication[37] and occur at exact harmonics of the first vibrational transition. They are definitely not due to a Raman transition, induced by the intensity at ω_L, while the molecule undergoes a transition with a change of two in the vibrational quantum number. The matrix element for such a transition is quite small and the anharmonicity of the vibration is sufficiently large that the frequency of spontaneous Raman emission for the double vibrational quantum transition $\omega_L - \omega_{v=0\to 2}$ does not coincide with $\omega_L - 2\omega_{v=0\to 1}$.

The theory of section 4-5 can be applied immediately. The Raman susceptibility is given by Eq. (4-70) in a very simple form that is identical to one used by Terhune[39] et al. He has an additional factor of $\frac{1}{4}$. Our Raman susceptibility is four times larger because of a redefinition of the field amplitudes. The expression for the gain is given by Eq. (4-68). For a given laser power our value of $|E_L|^2$ is four times smaller so that the stokes gain per centimeter comes out to be the same as calculated by Terhune for the same power level of the laser. Eq. (4-70) can be generalized slightly by assuming that the excited vibrational state may also have some partial population according to the Maxwell-Boltzmann distribution. The expressions for the stokes susceptibility and gain will here be rewritten in the form

$$-\chi_s'' = \frac{N\{1 - \exp(-\hbar\omega_v/kT)\}(\eta\omega_v\alpha_0/\omega_L - \omega_{ng})^2}{\hbar\Gamma_v} \tag{5-29}$$

The power gain per centimeter is twice the amplitude gain given by (4-68),

$$g_{\text{stokes}} = \frac{8\pi^2(-\chi_s'')|E_L|^2}{n\lambda_0} \tag{5-30}$$

The vacuum wave length λ_0 and the index of refraction n of the medium have been introduced. The difference between k_{sz} and k_s has been ignored in Eq. (5-30). The power flux density in the laser beam is given by $(nc/2\pi)|E_L|^2$.

A reasonable value for the Raman polarizability of a free

molecule $(\eta\omega_v\alpha_0/\omega_L - \omega_{ng}) \sim 10^{-25}$ esu. This quantity is of course directly related to the incoherent Raman scattering cross section as discussed in section 2-8. The Raman polarizability is an order of magnitude smaller than the polarizability caused by a pure electronic transition, not involving the nuclear vibrational coordinates. With a typical value for the half width at half maximum of 3 cm^{-1} for the width of the vibrational transition or $\Gamma_v = 2\pi \times 10^{11}$ and $N = 10^{22}$, one finds $|\chi_S''| \sim 1.6 \times 10^{-13}$ esu. When the Lorentz correction factor of $(\epsilon_L + 2)^2(\epsilon_S + 2)^2/81$ is applied, this becomes 6.5×10^{-13} esu. This value of $|\chi_S''|$ is of course about an order of magnitude larger than the nonresonant susceptibilities in the same order of nonlinearity of section 5-8. One gains a factor $\omega_v/\Gamma_v \sim 10^3$ on the vibrational resonance, but looses a factor $10 - 100$ on the electron-nuclear vibration coupling. Eq. (5-30) gives the stokes power gain coefficient for a laser flux density of 2 megawatt/cm^2, corresponding to $|E_L| = 53$ esu, $g_{stokes} = 1.4 \times 10^{-3}$ cm^{-1}. The stokes gain for a single traversal of a 10 cm long cell would be $e^{0.014}$. This would require about 99% reflectivity of the mirrors to insure oscillation. The power level inside the laser cavity is probably an order of magnitude higher, or 20 megawatt. In this case the gain on a cell traversal would be $e^{0.14}$ and Raman laser action is readily accomplished.

Our numerical example is in reasonable agreement with the figure quoted by Hellwarth for nitrobenzene.[40] He estimates a gain coefficient of 0.3 cm^{-1} at a pump level of 100 megawatt/cm^2, corresponding to a photon flux of 3×10^{26}/cm^2 sec. The total Raman scattering cross section, integrated over $\Omega = 4\pi$, was measured to be 6×10^{-30} cm^2. The vibrational energy is 1345 cm^{-1} with a half width at half intensity of 4 cm^{-1}. The value of $|\chi_S''|$ for nitrobenzene would thus be 2.4×10^{-12} esu. The precision of such determinations is subject to large uncertainties. All the problems discussed for the lower order nonlinearity in section (5-2) reappear in an aggravated form in the Raman effect. Measurements of the Raman scattering cross section in benzene, with $\omega_v/2\pi c = 992$ cm^{-1} and $\Gamma_v/2\pi c = 3.1$ cm^{-1} yield a Raman susceptibility $(4.8 \pm 1.2) \times 10^{-12}$ esu. This number is in reasonable agreement with the observed threshold for a benzene Raman laser.

Apart from the peculiar dip in the gain constant near the phase-matched direction with the antistokes radiation shown in Figure 4-13, the Raman gain should be nearly isotropic. It is indeed possible to obtain the stimulated stokes radiation at an angle to the laser pump beam.[41] The interaction volume in the parallel arrangement of beams is of course usually much larger. Since the effective length in the off-axis direction is shorter, much effort must be made to minimize stokes reflection coefficients in the forward direction and to have highly reflecting mirrors in the off-axis direction. If a laser beam of 1 cm length is focused by a cylindrical lens, one obtains a path of 1 cm for stokes radiation with high gain, at right angles to the original laser beam. With mirrors placed for feedback along this path, Tennenwald[42] has achieved stokes oscillation at right angles to the pump.

The oscillation starts, in general, from the noise. It is easiest to consider the spontaneous Raman scattering from a slab of material, normal to the direction of stokes propagation, since the input signal for the stokes amplifier is represented by adjacent slabs. The considerations are entirely analogous to the build-up of oscillations in traveling wave tubes. The oscillation condition may in general be written in the "Barkhausen" form for small signal oscillation,

$$r_1 r_2 \exp\{2i(k_S + \Delta\kappa)d\} = 1 \qquad (5\text{-}31)$$

Here r_1 and r_2 are the complex amplitude reflection coefficients at ω_S of the two mirrors, d is the length of the cell, and $\Delta\kappa$ is the complex root, given by Eqs. (4-66) or (4-67). The condition expresses the fact that the amplitude gain on a closed loop is equal to unity for self-sustained oscillation. The imaginary part of the condition is equivalent to the Townes-Schawlow condition for laser oscillation. It determines the threshold value for $|E_L|^2$. If this value is initially higher, the stokes amplitude will build up to such a level that the additional loss decreases the pump level to a self-consistent value. In fact, two oscillation conditions, for ω_L and ω_S, should be satisfied simultaneously in the steady state. The real part of the condition Eq. (5-31) determines the frequency of oscillation. For the rather broad natural width of the Raman lines this precise frequency is determined by the mirror spacing d. If the mirrors are not at the ends of the cell, the expression on the left has of course to be multiplied with propagation factors for the other media in the resonator.

The stokes intensity can build up to serve as the pump of the second stokes line, and so on. Oscillation could be prevented at these new frequencies by selective antireflective coatings or specific absorbers in the Raman cell. Without such precautions, the output power will spill over into more and more stokes components of comparable intensity as the laser pump power is increased. More than sixty per cent of output power at stokes frequencies has been observed in a Raman laser. This percentage should probably be higher were it not for the fact that the laser beam is usually far from a homogeneous plane wave. Sections of the beam where the intensity is lower are only converted partially or not at all. From the standpoint of conversion, a ruby laser with a few very bright filaments would be better than a homogeneous cross section with the same total power flux. The variation of intensity with time during the pulse is also important. It is possible to obtain peak powers of 500 megawatts or more in unfocused laser beams amplified by a laser amplifier. In a 10 cm long path in nitrobenzene the gain would be e^{15}. Clearly a feedback by accidental scattering of one part per million would be sufficient to set the system into oscillation. In a path of 30 cm the gain would be e^{45}. This means that a single photon at the input would on the average have created 10^{20} photons. This number is, however, larger than the initial number of laser quanta in the pulse, which is less than 10^{18}. The laser beam is exhausted before it reaches the end of the cell. If the beam is 1 cm in diameter, appreciable stokes gain occurs up to angles of several degrees. This stokes radiation just off the forward direction is observable say after passage through about 15 cm in the cell. Due to the exponential character of the gain the emission appears to set in at a rather well-defined position in the cell. It may be considered as amplified noise. If the cell is shorter, a minimum of feedback is necessary. Even in cells with Brewster angle windows sufficient back scattering of say $1:10^6$ is present. The off-axis stokes radiation then appears to come from the last few millimeters near the cell end, as those rays have had the highest gain. In short cells, say < 10 cm, transit time effects can be neglected. The gain in the backward direction should be the same as in the parallel forward direction. Preliminary experimental data indicate the backward intensity to be a factor of two to four less. If the damping constant changes with direction by a few per cent, as suggested in section 4-5, this would reduce a forward gain of e^{20} to a backward gain of e^{19}, enough to account for the effect.

5-10 HIGHER ORDER STOKES AND ANTISTOKES RADIATION

Power flux densities in excess of 1000 megawatts/cm^2 are readily obtained in the focus of an external laser beam. Terhune[43] was the first to use the arrangement shown in Figure 1-4b. He observed visible colored rings in addition to copious stokes radiation in various orders from benzene, liquid N_2, and several other liquids. Soon afterwards similar observations have been reported by several other workers.[44,45]

A typical distribution of intensity as a function of frequency and direction is shown in Figure 5-9. By inserting a frosted glass plate immediately behind the Raman cell, the frequency spectrum integrated over all directions can be measured by imaging the plate on the slit of a high-resolution instrument. Stoicheff[44] has described some peculiar features about the width and structure of the lines. At very high power levels a broadening occurs which may amount to 50 cm^{-1}, or ten times the width of the spontaneous Raman line. Just

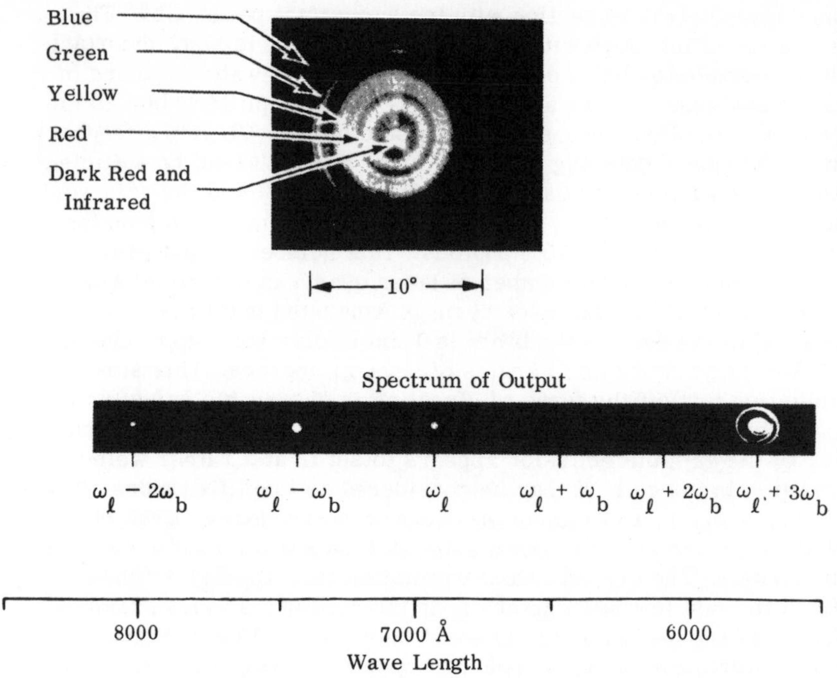

Figure 5-9. Distribution in direction and in frequency of stokes and antistokes light, generated at the focus of a pulsed ruby laser in benzene (after Terhune).[17]

above threshold the antistokes lines can be considerably narrower than Γ, or $0.2\ \mathrm{cm}^{-1}$.

Terhune and coworkers[46] have observed very strong antistokes light in the forward direction from gases. The observed spectrum from H_2 is shown in Figure 5-10. The light in the forward direction gives a "white" impression, because all antistokes cones overlap. The effect in gases is strong in spite of the relatively smaller number of molecules per unit volume in a given vibrational-rotational state, because the line width Γ is very small. At pressures above one atmosphere the line width is inversely proportional to the pressure. This means according to Eq. (5-29) that the gain per unit length is independent of the gas pressure as observed experimentally. At lower pressures Γ is determined by the Doppler width. Before that point is reached, however, the spectral distribution in the laser output, usually $0.1\ \mathrm{cm}^{-1}$ wide, becomes comparable in width to Γ. In that case the gain ceases to be independent of pressure for a given total laser power. Not all components or modes in the laser

Wave Length	3721 Å	4402 Å	5388 Å	6943 Å	9755 Å
Relative Energy	0.2%	2%	2%	70%	25%

Spectra from Gaseous H_2

Figure 5-10. Spectrum of coherent Raman effect, generated at the focus of a pulsed ruby laser in hydrogen gas. The exposure is from one pulse, lasting 2×10^{-8} sec (after Minck et al.).[46]

beam can contribute to the stimulation of a given stokes mode. In that case the convolution of the Lorentzian vibrational line shape and the laser spectral distribution must be taken.

Terhune[46] estimates that the gain in a single passage through the focus in H_2 gas should be e^{500}. For the geometry of the diffraction limited focus of section 5-1 the intensity in the focal area is increased by $d/\delta \propto f^{-2}$. The effective length of the focal volume is, however, decreased by f^{+2}. The Raman gain through the focus is independent of the f-number, and is, according to Eqs. (5-1), (5-2), and (5-30) given by

$$\ln(g_{SZ}) = \frac{8\pi^2(-\chi_S'')|E_L|^2}{n\lambda_0}\left(\frac{d}{\delta}\right)^2 \xi = \frac{8\pi^2(-\chi_S'')|E_L|^2}{n\lambda}\,\frac{2nd^2}{\lambda_0}$$

It is 3 db larger than the gain in an unfocused beam over a distance necessary to increase the beam diameter by two due to diffraction.

More recent data have revealed that even in liquids and solids there is considerable intensity at $\omega_a = \omega_L + \omega_v$ in the forward direction, in addition to the cone. The cone angles measured in liquids, as well as the spectral widths of the various stokes and antistokes lines, appear to depend on the details of the experimental arrangement and nature of the focus.

Stoicheff and Chiao[47] have accurately measured the apex angles of the cones in calcite. In this case the focus of the laser beam is not in the crystal itself, because it would be damaged at the extremely high intensity. All Raman light appeared to come from the end of the crystal as explained before. Stoicheff and Chiao also observed dark rings in the intensity distribution at the stokes frequencies in several orders and bright rings at the antistokes

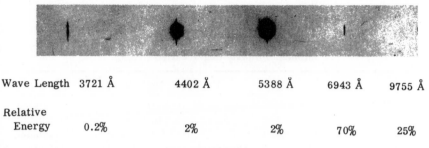

TABLE 5-3
Observed Angles of Emission of Antistokes Light in Various Orders, Stimulated by a Ruby Maser in Calcite. The Directions for Linear Momentum Matching are also Given.[47]

		Emission Angle in 10^{-2} Radian		
Frequency	Wave Length in Å	Observed	Extrapolated to Lenses for Infinite Focal Length	Calculated for Linear Momentum Matching
$\omega_L + \omega_v$	6456.0	2.89 ± 0.03	2.50 ± 0.03	2.49
$\omega_L + 2\omega_v$	6033.2	5.42 ± 0.08	5.03 ± 0.08	4.91
$\omega_L + 3\omega_v$	5662.3	7.80 ± 0.2	7.64 ± 0.2	7.29
$\omega_L + 4\omega_v$	5334.4	10.64 ± 0.4	10.2 ± 0.4	9.61

frequencies. The half apex angles of the various antistokes cones are listed in Table 5-3, together with the angle, calculated for exact momentum matching from the accurately known linear index of refraction.

The angular distribution of stokes intensity may be understood as follows. Although the gain per unit length is nearly isotropic, there will be a rapid drop at angles of a few degrees from the forward direction, because the effective interaction length with the laser beam is curtailed. Note that a very small change in path length, varying approximately as $(\cos \theta)^{-1}$ can produce a large change in observed intensity because of the exponential nature of the gain function. Apart from this geometric effect, there is a very sharp dip at the angle θ_0, corresponding to linear momentum matching for which $k_S^\circ + k_a^\circ = 2k_L^\circ$ shown in Figure 4-13. This explains the dark ring in the stokes background in calcite. In other materials this ring should also exist, but is presumably obscured by scattering processes.

The amplified wave according to section 4-6 has always some partial antistokes character. In the forward direction the gain is so high that partial antistokes character is observable. According to Eqs. (4-92) and (4-68) or (5-30), the ratio of antistokes to stokes intensity is $(\frac{1}{2}g_S)^2 \Delta k^{-2}$, or roughly the square of the amplitude gain over the coherence length. In the forward direction Δk^{-1} for nitrobenzene is about 10^{-2} cm. For a power gain of 10 db/cm the partial character would be 0.01 per cent, which appears to be the right order of magnitude.

Near the direction of linear momentum matching the gain constant decreases, but the partial antistokes character increases. In Figure 4-14 the antistokes intensity has been plotted as a function of Δk for two typical values of the gain in the forward direction. When the latter is 120 db, the antistokes is a maximum for a momentum mismatch $\Delta k = 2g_S$. This mismatch is not very sensitive to the overall gain. The deviation in the transverse component of momentum from the direction of momentum match may be determined from,

$$(k_x^\circ + \Delta k_x)^2 + (k_{zs}^\circ - \tfrac{1}{2}\Delta k)^2 = (k_x^\circ)^2 + (k_{zs}^\circ)^2 = (k_s^\circ)^2$$

For small relative variations in k_x one has

$$\Delta k_x = \frac{1}{2}\frac{k_{sz}^\circ}{k_x^\circ}\,\Delta k = \frac{\Delta k}{2\mathrm{tg}\,\theta_0}$$

where θ_0 is the exact phase-matched direction; $\theta_0 = 2.5^\circ$ for ruby light in nitrobenzene. The angle of the direction of antistokes radiation as observed outside the cell is given by $\sin\theta = (k_x^\circ + \Delta k_x)/k_s^{vac}$, where k_s^{vac} is the wave vector in vacuum. For small angles and small variations one has immediately $\Delta\theta/\theta_0 = \Delta k_x/k_x^\circ = \Delta k/2k_s^\circ \mathrm{tg}^2\theta_0$. The direction of maximum antistokes intensity is offset from the matching direction by

$$\Delta\theta = \theta_0 g_s/k_s^\circ \mathrm{tg}^2\,\theta_0 \approx g_s/k_s^\circ\,\theta_0 \qquad (5\text{-}32)$$

For $g_s = 4\ \mathrm{cm}^{-1}$ corresponding to a gain of 18 db/cm, one finds $\Delta\theta = 0.1^\circ$ for nitrobenzene. The gain in the focal region may be even larger with a corresponding larger value of $\Delta\theta$. The experimentally reported value for the angular direction of the first antistokes ring in nitrobenzene is 3.1° or $\Delta\theta = 0.6^\circ$. Other workers report other deviations. The results appear to be very sensitive to detailed experimental conditions. Chiao and Stoicheff report smaller deviations $\Delta\theta$ in calcite when lenses with longer focal length are used. The extrapolated experimental values for infinite focal length correspond to $\Delta\theta \sim 0.1^\circ$.

The theoretical predictions are not in disaccord with the experimental results, when it is kept in mind that the assumptions on which the theory of section 4-6 is based are not fulfilled in the actual experimental conditions. Depletion of the laser power occurs and higher stokes and antistokes combination frequencies are not suppressed. A most important factor is probably that the laser intensity is far from uniform.

It can be argued that the observed antistokes radiation should come predominantly through the region of highest laser intensity. Radiation passing through this region at a smaller angle than $\theta_0 + \Delta\theta_{max}$ will not be amplified much, as a glance at Figure 4-14 shows. Radiation passing at a larger angle will still have an appreciable gain. The nature of the gain curves in Figure 4-14 is such that the antistokes intensity decreases rapidly at angles smaller than $\theta_0 + \Delta\theta_{max}$ where $\Delta\theta_{max}$ is the angular deviation appropriate to the region of maximum gain.

For gases the phase-matched direction is essentially the forward direction, $k_x^\circ = 0$. In this case the maximum gain for antistokes occurs at an angle

$$\theta_{max} \approx \sin\theta_{max} = \frac{\Delta k_x}{k_s^\circ} = \frac{(\Delta k\, k_{sz}^\circ)^{1/2}}{k_s^\circ} \approx \left(\frac{2g_s}{k_s^\circ}\right)^{1/2}$$

For $g_s = 4$ corresponding to a gain of 18 db/cm, this angle would be 8×10^{-3} radian or about 0.5°. It would increase as the square root of the gain constant. The antistokes light in a gas comes out essentially, but not quite, in the forward direction. The stokes gain should also have a dip close to the forward direction. This may be obscured by scattering and by geometrical factors.

Further frequencies can now be generated in a variety of ways as explained by Terhune[39] and Garmire, Pandarese, and Townes.[48] The higher stokes lines in the forward direction, $\omega_{ss} = \omega_L - 2\omega_v$, $\omega_{sss} = \omega_L - 3\omega_v$, etc., by a cascade of Raman processes has already been mentioned. There is, however, also a dark ring in ω_{ss}, characterized by a wave vector $\mathbf{k}_{ss}$. It is caused by the parametric coupling of four waves. The antistokes wave with momentum $\mathbf{k}_a^{max}$ in the direction of the bright ring, $\mathbf{k}_L$ the laser momentum, and a stokes wave in an appropriate direction chosen out of the forward stokes cone in such a manner that $\mathbf{k}_a - \mathbf{k}_L = \mathbf{k}_s - \mathbf{k}_{ss}$. The left-hand side is completely fixed in magnitude and direction. The magnitudes of the wave vectors on the right are known from the linear dispersive properties of the medium. Thus the direction of $\mathbf{k}_{ss}$ can be found. This ring is created by a nonlinear source polarization

$$\mathbf{P}^{NLS}(\omega_{ss}) = \chi E_a^* E_L E_s \exp\{i(\mathbf{k}_L + \mathbf{k}_s - \mathbf{k}_a)\cdot\mathbf{r} - i\omega_{ss}t\}$$

In a similar manner a bright second order antistokes ring is expected from the source,

$$P^{NLS}(\omega_{aa}) = \chi E_a E_L E_s^* \exp\{i(\mathbf{k}_a + \mathbf{k}_L - \mathbf{k}_s)\cdot\mathbf{r} - i\omega_{aa}t\}$$

This process may be repeated, for example, a ring at $\omega_{aaa} = \omega_L + 3\omega_v$ will be created by the source $P^{NLS}(\omega_{aaa}) = \chi E_{aa} E_L E_s^* \times \exp\{i(\mathbf{k}_{aa} + \mathbf{k}_L - \mathbf{k}_s)\cdot\mathbf{r} - i\omega_{aaa}t\}$ in a direction of $\mathbf{k}_{aaa}$ and $\mathbf{k}_s$ for which the momentum match $\mathbf{k}_{aaa} + \mathbf{k}_s = \mathbf{k}_{aa} + \mathbf{k}_L$ is satisfied.

Experimentally five or six antistokes rings have been observed and a similar number of higher order stokes rings. This implies that the waves are intense enough to beat together many times. This may offer a clue to the large width, five to ten times the spontaneous width, of the components created in such intense light fields. The gain curves in Figure 4-14 and the discussion in section 4-6 show that there is appreciable gain over a range of frequencies of the order of half the line width. The Fourier components at ω_{aa}, in addition to creating a polarization at ω_{aaa}, can also beat back with ω_L and components of ω_s to recreate ω_a. The width of this recreated intensity will, however, be the convolution of the original width with itself. If this convolution is repeated five times, the width of the lines may reach 5Γ or more. This argument is qualitative. The waves are so intense and interact so strongly that a complete simultaneous solution of all coupled waves would be necessary. The parametric arguments can only give a qualitative description of the situation. The detailed position of higher order dark and bright rings can only be obtained from the coupled wave approach.

5-11 RAMAN-TYPE SUSCEPTIBILITIES

The creation of the stokes and antistokes waves in the focus of a high-intensity laser beam is a fascinating phenomenon, but the experimental conditions are not sufficiently well defined to provide a test of the theory, nor can they unravel the various physical mechanisms. The situation may be compared with the understanding of the operation and characteristics of a vacuum tube. One will study first the behavior of the tube as a small signal amplifier, rather than a high-level oscillator. In this spirit the properties of Raman laser media should be studied in thin cells, which are not capable of generation of coherent Raman processes by themselves under the most intense laser illumination at ω_L. When a small signal, either at ω_s or ω_a, is also incident on the cell, the gain can be measured. Experimentally one could always keep the gain less than a factor of 2 or 3. There would be no depletion of the laser intensity, no significant creation of higher order stokes and antistokes lines. The intensity, polarization, direction, and frequency of laser beam and the stokes beam would be independently controlled. Ideally, each of the beams would contain only one mode, i.e., they should be monochromatic and diffraction limited. Such experiments would be capable of yielding reliable values for the Raman-type susceptibilities and to test the details of the theory presented in Chapters 2 and 4. A possible experimental arrangement is shown in Figure 5-11.

A powerful pulsed laser beam is split and one fraction generates stokes radiation in a Raman cell. The laser light is filtered out. The

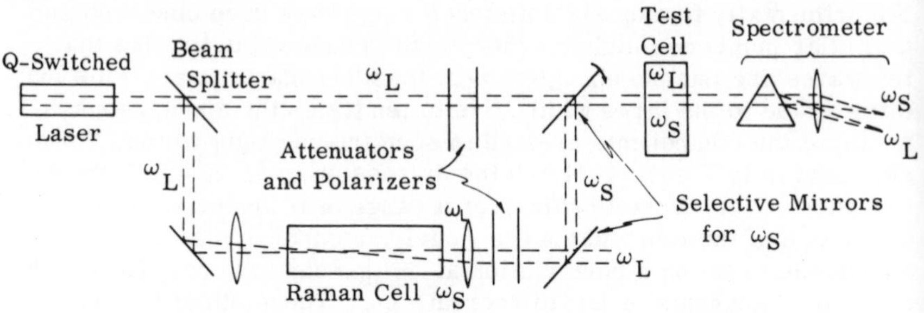

Figure 5-11. Experimental arrangement for the measurement of Raman-type susceptibilities. The stokes gain can be determined as a function of intensity direction and polarization, of both the laser and the stokes beam, in a short sample.

stokes light is, after suitable attenuation and polarization, recombined with the other part of the laser beam in a variable direction. The two beams then traverse the thin sample cell. The gain can thus be measured as a function of direction, polarization, intensity. The generation of antistokes radiation near the phase-matched direction can also be measured.

Similar experiments have already been carried out by Bret and Mayer,[49] although the direction between the stokes and antistokes beam has not been varied. First ω_L and ω_S are both created in a Raman laser. Then the two beams are individually attenuated by selective filters. They are polarized and the angle between their polarization can be varied by passing through a retardation plate of prescribed thickness. Mayer[50] has in this manner observed that the stokes gain in benzene for the polarization perpendicular to the laser polarization is vanishingly small compared to the gain for the parallel polarization. This does not imply that the tensor element χ_{xxyy} is very small compared to χ_{xxxx}. The exponential character of the gain is such that a ratio of 1.5 or 2 between these elements is sufficient to give a very large ratio for the observed intensities of the two possible stokes polarizations. He has also found important saturation effects on the gain, i.e., the gain factor depends on the intensity of incident stokes signal to be amplified.

Terhune[50] has measured the parametric generation of light at $2\omega_L - (\omega_L - \omega_v) = \omega_L + \omega_v$ in thin samples of a variety of cubic crystals and liquids. A focused beam from a benzene Raman laser contains linear polarized light at ω_L and $\omega_L - \omega_v$ with $\omega_v/2\pi c =$ 992 cm^{-1}. All other frequencies have been filtered out of the beam.

The light propagates along [001] in the platelets of the cubic crystals. The angle between the [100] axis and the parallel polarizations at ω_L and $\omega_L - \omega_V$ is denoted by θ. For $\theta = 45°$ a different intensity at $\omega_L + \omega_V$ is created than at $\theta = 0$ or $90°$, indicating the anisotropy of the fourth-rank susceptibility tensor in cubic crystals.

The parametric generation shows the typical oscillatory behavior as a function of the thickness of the crystals. The coherence length of the three waves all propagating in the z-direction is of the order of 1 mm. The relative yield of light at $\omega_L + \omega_V$ is a measure for the fourth-rank susceptibility. They are measured relative to the yield in benzene itself. The effect in benzene is of course largest because of the resonance of the Raman-type susceptibility. In the other substances essentially the nonresonant part is measured. The numerical values for the susceptibility $\chi_{xxxx}\{\omega = -(\omega_L - \omega_{v,benzene})$ $+ \omega_L + \omega_L\}$ are listed in the second column of Table 5-4 for various cubic crystals and liquids. In the third column the anisotropy ratio

TABLE 5-4

The Susceptibility $\chi_{xxxx}\{\omega = -(\omega_L - \omega_v) + \omega_L + \omega_L$ in a Number of Cubic Crystals and Liquids, for ω_L Equal to the Ruby Laser Frequency and $\omega_V/2\pi c = 992$ cm^{-1}. The Anisotropy Factor is Shown in the Third Column. The Phase Coherence Length for the Parametric Generation of $\omega_L + \omega_v$ is Given in the Fourth Column. (Courtesy of R. W. Terhune)

Material	$\|\chi_{xxxx}\|$ in 10^{-13} erg^{-1}cm^3	$\dfrac{\chi_{xxxx}}{\chi_{xyxy} + \chi_{xxyy}}$	ℓ_{coh} in mm
LiF	0.24	+ 0.74 ± 0.10	3.27
KI	0.29	0.54	0.45
CaF$_2$	0.4	0.66	2.94
MgO	1.2	0.61	1.27
KCl	2.3	1.10	0.92
NaCl	2.1	0.78	0.88
KBr	3.6	0.91	0.67
Fused SiO$_2$	0.8	1	1.45
C$_6$H$_5$CH$_3$	4.8	1	0.80
C$_6$H$_5$Br	7.2	1	0.65
C$_6$H$_6$	48[a]	1	0.70

[a] This resonant value is calculated from the Raman scattering cross section. It is used as a calibration for the other, nonresonant, values in this column.

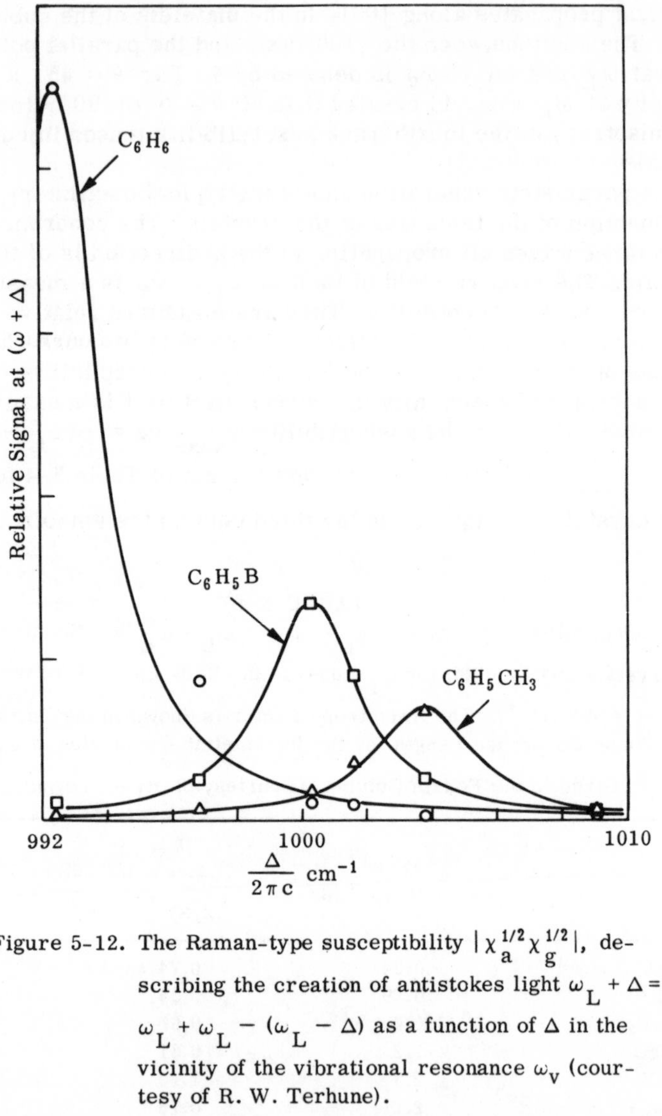

Figure 5-12. The Raman-type susceptibility $|\chi_a^{1/2}\chi_g^{1/2}|$, describing the creation of antistokes light $\omega_L + \Delta = \omega_L + \omega_L - (\omega_L - \Delta)$ as a function of Δ in the vicinity of the vibrational resonance ω_v (courtesy of R. W. Terhune).

$(\chi_{xxxx}/\chi_{xyxy} + \chi_{xxyy})$ is listed (compare section 5-8). The coherence length in the last column is computed from the known linear dispersion. The Raman susceptibility for benzene derived from the Raman scattering cross section was used as a reference, since absolute determinations of the parametric yield are fraught with difficulties mentioned previously.

Terhune[50] was also able to reproduce a resonance curve for the

Raman-type susceptibility by utilizing Raman lasers and sample cells with a number of liquids with nearly equal vibrational resonant frequencies. He used the vibrations $\Delta/2\pi c$ = 992.4 cm^{-1} in benzene, 996.8 cm^{-1} in analine, 1000.3 cm^{-1} in bromobenzene, 1001.6 cm^{-1} in benzonitrile, 1003.8 cm^{-1} in toluene, and 1009.1 in fluorobenzene as generators. The results of the signal at $\omega_L + \Delta$ in benzene, bromobenzene, and toluene are shown in Figure 5-12. He also used deuterated benzene with $\Delta/2\pi c$ = 945 cm^{-1} to obtain a measurement for the nonresonant part of the Raman susceptibility. Terhune found $\chi^{NR}/(\chi_a \chi_s)^{1/2}$ = 0.05 for benzene, 0.25 for bromobenzene, and 0.16 for toluene. The curves in Figure 5-12 should show a somewhat greater asymmetry to be consistent with these determinations.

Clearly, a large amount of careful work remains to be done in the quantitative determination of complex Raman susceptibilities, their tensorial symmetry and dispersion properties in solids, liquids, and gases. A beginning has, however, been made and further progress may be expected in the immediate future.

REFERENCES

1. P. A. Franken, A. E. Hill, C. W. Peters, G. Weinreich, *Phys. Rev. Letters*, **7**, 118 (1961).
2. J. A. Giordmaine, *Phys. Rev. Letters*, **8**, 19 (1962).
3. P. D. Maker, R. W. Terhune, M. Nisenoff, C. M. Savage, *Phys. Rev. Letters*, **8**, 21 (1962).
4. R. W. Terhune, P. D. Maker, and C. M. Savage, *App. Phys. Letters*, **2**, 54 (1963).
5. M. Born and E. Wolf, *Principles of Optics*, Pergamon Press, New York (1959), p. 434 ff.
6. D. A. Kleinman, *Phys. Rev.*, **128**, 1761 (1962).
7. J. Ducuing and N. Bloembergen, *Phys. Rev. Letters*, **10**, 474 (1963).
8. M. Bass, P. A. Franken, A. E. Hill, C. W. Peters, G. Weinreich, *Phys. Rev. Letters*, **8**, 18 (1962).
9. A. W. Smith and N. Braslau, *IBM Journal Res*, **6**, 361 (1962).
10. J. F. Ward and P. A. Franken, *Phys. Rev.*, **133**, A 183 (1964).
11. I. P. Kaminow, *Phys. Rev. Letters*, **6**, 528 (1961); W. W. Rigrod and I. P. Kaminow, *Proc. IEEE*, **51**, 137 (1963).
12. K. E. Niebuhr, *App. Pnys. Letters*, **2**, 136 (1963).
13. A. Ashkin, G. D. Boyd, and J. M. Dziedzic, *Phys. Rev. Letters*, **11**, 14 (1963).
14. J. Ducuing and N. Bloembergen, *Phys. Rev.*, **133**, 1493 (1964).
15. A. T. Forrester, R. A. Gudmundsen, and P. O. Johnson, *Phys. Rev.*, **99**, 1961 (1955).

16. R. C. Miller, D. A. Kleinman, and A. Savage, *Phys. Rev. Letters*, **11**, 146 (1963).
17. J. P. van der Ziel and N. Bloembergen, *Phys. Rev.*, **135**, A 1662 (1964).
18. R. C. Miller, *Phys. Rev.*, **131**, 95 (1963), *Phys. Rev.*, **134**, A1313 (1964), *App. Phys. Letters*, **5**, 17 (1964).
19. R. A. Soref and H. W. Moos, *J. App. Phys.*, **35**, 2152 (1964).
20. N. Bloembergen, R. K. Chang, J. Ducuing, and P. Lallemand, *Proceedings International Conference on Semiconductors*, Dunod, Paris, July 1964.
21. A. Savage, to be published.
22. D. A. Kleinman, *Phys. Rev.*, **126**, 1977 (1962).
23. H. R. Phillip and H. Ehrenreich, *Phys. Rev.*, **129**, 1550 (1963).
24. B. Lax, J. G. Mavroides, and D. F. Edwards, *Phys. Rev. Letters*, **8**, 166 (1962).
25. R. W. Terhune, P. Maker, and C. M. Savage, *Phys. Rev. Letters*, **8**, 21 (1962).
26. P. S. Pershan, *Phys. Rev.*, **130**, 919 (1963).
27. J. A. Giordmaine, *Proceedings Third Conference on Quantum Electronics*, Paris, 1963, ed. P. Grivet and N. Bloembergen, Columbia University Press, New York, 1964, p. 1449.
28. P. D. Maker, R. W. Terhune, and C. M. Savage, Ibid., p. 1559.
29. D. A. Kleinman, *Phys. Rev.*, **125**, 87 (1962).
30. R. Braunstein, *Phys. Rev.*, **125**, 475 (1962).
31. W. Kaiser and C. G. B. Garrett, *Phys. Rev. Letters*, **7**, 229 (1961).
32. J. J. Hopfield, J. M. Worlock, and K. Park, *Phys. Rev. Letters*, **11**, 414 (1963).
33. S. Singh and L. T. Bradley, *Phys. Rev. Letters*, **12**, 612 (1964).
34. P. D. Maker, R. W. Terhune, and C. M. Savage, *Phys. Rev. Letters*, **12**, 507 (1964).
35. J. A. Giordmaine and J. A. Howe, *Phys. Rev. Letters*, **11**, 207 (1963).
36. E. J. Woodbury and W. K. Ng, *Proc. IRE*, **50**, 2367 (1962).
37. G. Eckhardt, R. W. Hellwarth, F. J. McClung, S. E. Schwarz, D. Weiner, and E. J. Woodbury, *Phys. Rev. Letters*, **9**, 455 (1962).
38. G. Eckhardt, D. P. Bortfeld, and M. Geller, *App. Phys. Letters*, **3**, 36 (1963).
39. R. W. Terhune, *Solid State Design*, **4**, 38, November 1963.
40. R. W. Hellwarth, *Applied Optics*, **2**, 847 (1963).
41. H. Takuma and D. A. Jennings, *App. Phys. Letters*, **4**, 185 (1964).
42. J. H. Dennis and P. E. Tannenwald, *App. Phys. Letters*, **5**, 58, (1964).
43. R. W. Terhune, see ref. Chap. 4, nos. 17 and 24.

44. B. P. Stoicheff, *Phys. Letters*, **7**, 186 (1963).
45. H. J. Zeiger, P. E. Tannenwald, S. Kern, and R. Herendeen, *Phys. Rev. Letters*, **11**, 419 (1963).
46. R. W. Minck, R. W. Terhune, and W. G. Rado, *App. Phys. Letters*, **3**, 181 (1963).
47. R. Chiao and B. P. Stoicheff, *Phys. Rev. Letters*, **12**, 290 (1964).
48. E. Garmire, F. Pandarese, and C. H. Townes, *Phys. Rev. Letters*, **11**, 160 (1963).
49. G. Bret and G. Mayer, Comptes Rendues, **258**, 3265 (1964).
50. R. W. Terhune, *Study of Optical Effects Due to a Polarization of Third Order in the Electric Field Strength.* The author is indebted to Dr. Terhune for making this information available before publication.

6

CONCLUSION

There are many other topics that should properly be included under the broad title of nonlinear optics. The brevity of this chapter does no justice to them.

The photoelectric effect is essentially a phenomenon with a quadratic response function; the detected photocurrent is proportional to the square of the field strength. It is of considerable interest to know what the frequency response is. If the two light waves at ω_1 and ω_2 fall on a photosensitive material, what is the component in the photocurrent at $\omega_1 - \omega_2$? The homogeneous damping time of the excited electronic state τ is the determining factor and the response behaves as $\{1 + (\omega_1 - \omega_2)^2 \tau^2\}^{-1/2}$ for a wide class of photoelectric and photo conductive phenomena.[1,2,3]

Mixing experiments of light with microwaves have not been discussed, although they give rise to many important devices. The nonlinearities of plasmas[4,5] in the microwave and far-infrared region ar omitted.

Intensity dependent absorption is described by Eq. (2-69). A solution of the wave equation with this power dependent absorption would show how the attenuation proceeds when a high-intensity wave passes through a medium that can be saturated. The process has an important application in saturable filters. They become transparent at high intensities and may be used as switching elements in pulsed lasers.[6]

Other phenomena well known at microwave and radio frequencies from magnetic resonance may have their counterpart at optical frequencies. One may consider 90° and 180° pulses by light of very high intensity acting on sufficiently narrow energy levels. There will be intensity dependent shifts of the resonant frequency, etc.

Important theoretical questions about the interactions of light with light in vacuum, the nonlinear interaction of light with a relativistic electron, quantum fluctuations, and quantum noise in parametric and Raman devices are also omitted. The only excuse is that the treatment of these questions would carry us too far from the central theme of this monograph, the nonlinear susceptibilities of matter.

6-1 NONLINEARITIES IN LASERS

This excuse does not apply to the omission of the role of nonlinearities in the operation of lasers. The behavior of any oscillator is determined by the nonlinearities in the characteristic. The amplitude of the oscillation is limited by the incipient saturation of the population in the levels between which the laser transition takes place. The rate equations for the populations in various types of lasers are well known. They correspond to the solution of the diagonal elements of the density matrix, according to the methods outlined in Chapter 2. There is no need to repeat them here. These considerations are, however, only valid for atoms or molecules at rest. They can be applied in a qualitative way to gas lasers, by treating the Doppler broadening in an ad hoc manner.

A correct description of the theory of the gas laser in which the density matrix also depends on the velocity and position of the atom has recently been published by Lamb.[7] The same question has also been treated in a series of papers by Haken and Sauermann.[8] When the laser oscillates in two modes, these modes may compete for the same atoms and interesting pulling effects occur. When the gas laser oscillates simultaneously in three longitudinal modes, with equal or almost equal spacing, a nonlinear coupling effect occurs. It is described by the nonlinear polarization produced by waves at ω_1 and ω_2 at the frequency $\omega_3 = 2\omega_2 - \omega_1$ and vice versa. If the free running oscillator frequency ω_3' is very close to ω_3, a pulling and lock-in of the oscillating frequencies has been observed.[9] Lamb has developed the theory for these nonlinearities in detail. He has also discussed the stability of oscillation conditions for the three coupled oscillator modes. In our discussion the traveling wave modes have been emphasized. With the exception of ring lasers, the mode pattern in oscillators is usually a standing wave pattern. Each standing wave can of course be considered as two traveling waves, but the complexity of the problem is greatly increased by this doubling of the number of waves.

The harmonic generation or Raman effect inside gas lasers is not very important at present power levels, although this question must be reexamined if one admits a focal point inside the resonator. Harmonic generation in semiconductor lasers (GaAs) is well known.[10,11] It can be calculated from the standard theory for a known parametric source distribution.

6-2 OTHER GEOMETRIES

The emphasis has almost exclusively been on plane waves of infinite cross section and plane boundaries of infinite dimensions. This approach is in keeping with the spirit of geometrical optics, where the behavior of pencil rays of light is used to describe more complex situations. The image formation by a lens, e.g., is derived in terms of Snell's law obtained from refraction of a plane wave at a plane boundary. In a similar manner the inhomogeneous source waves and their interference with the homogeneous waves may be analyzed in the passage of light through nonlinear lenses, prisms, multiple layers etc. The effects of finite cross section of the beam can be taken into account by diffraction according to Kirchoff's or Huygens' principle.

The diffraction theory should be extended to nonlinear media. In particular, the growth of a wave in the focal region of a lens should be considered more precisely. It is, of course, straightforward to derive a parametric solution with a prescribed distribution of the nonlinear source. Harmonic generation from little spheres or cylinders presents no special problem.

All equations have been derived for a steady state. It is true that the steady state is quickly established and probably exists even in a Q-switched laser pulse, lasting only 10^{-8} sec. Some further scrutiny of the problem of transients is desirable. Kroll[12] has recently investigated the transient problem for the case of stimulated Brillouin scattering. It appears that a steady state may not have time to develop during the Q-switched laser pulse due to the relatively slow velocity of the sound waves. In the case of Raman processes and parametric processes involving only light waves, the steady state solutions appear to be a good approximation to the physical situation.

A problem of great importance is that of the strength of materials under extremely intense fields. There are many indications that this is not so much a problem of optics, as one of heat deposition. At sufficiently high intensity some nonlinear dissipative process will set in. It may be Raman or Brillouin scattering, depositing vibrational energy, or it may be multiple photon absorption. In a crystal the process may well start near an imperfection or impurity. This energy is of course, deposited in a very short time in the curious geometry determined by the intensity distribution in the focus. It will thus lead quickly to a problem of shock waves in fluid mechanics.

6-3 CONCLUSION

This enumeration of unfinished business is undoubtedly not complete. A framework has, however, been established, within which these and other problems can be solved. First, the basic nonlinear optical properties of materials can be described in terms of existing classical or quantum mechanical concepts about the structure of matter. Second, these nonlinear properties can be described by complex susceptibility tensors of ascending rank. Third, these nonlinearities can, even at the highest attainable flux densities, be considered as small perturbations, and they can be incorporated in Maxwell's equations to give a set of coupled wave equations. Fourth, approximate solutions to these equations can be found by generalizing the methods of parametric traveling wave amplifiers. Fifth, this leads to a generalization of the well known laws of optical reflection, Snell's law, Fresnel's laws, etc. to the nonlinear case. Sixth, the method can be extended to include the interactions with vibrational waves. The coherent Brillouin and Raman effect can be treated along the same lines and expressed in terms of susceptibilities. Seventh, the measurement of these nonlinear susceptibilities, their tensorial symmetry and dispersive properties as a function of several frequencies, gives additional information about the structure of matter. Finally, a knowledge of these material properties will lead to the design of many nonlinear devices that will perform the same functions in the ultraviolet and infrared visible part of the electromagnetic spectrum that nonlinear tubes and circuit elements are known to perform at lower frequencies.

REFERENCES

1. P. S. Pershan and N. Bloembergen, *App. Phys. Letters*, **2**, 117 (1963).
2. G. J. Lasher and A. H. Nethercott, *J. App. Phys.*, **34**, 2122 (1963).
3. O. Svelto, P. D. Coleman, M. Di Domenico, Jr., and R. H. Pantell, *J. App. Phys.*, **34**, 3182 (1963).
4. P. M. Platzman and N. Tzoar, *Phys. Rev. Letters*, **12**, 573 (1964).
5. N. M. Kroll, A. Rou, and N. Rostoker, *Phys. Rev. Letters*, **13**, 83 (1964).
6. G. Bret and F. Gires, *Comptes Rendues* (Paris), **258**, 4702 (1964).
7. W. E. Lamb, Jr., *Phys. Rev.*, **134**, A 1420 (1964).
8. H. Haken and H. Sauermann, *Z. Physik*, **173**, 261 (1963); **176**, 47 (1963).
9. W. J. Witteman and J. Haisma, *Phys. Rev. Letters*, **12**, 617 (1964).
10. J. A. Armstrong, N. I. Nathan, and A. W. Smith, *App. Phys. Letters*, **3**, 68 (1963).
11. L. D. Malmstrom, J. J. Schlickman, and R. H. Kingston, *J. App. Phys.*, **35**, 248 (1964).
12. N. M. Kroll, *J. App. Phys.*, to be published.

Epilogue

Although no attempt has been made to update the material nor to correct all minor misprints and errors in the reprinted original edition, the following notes will elliminate some difficulties which past readers have experienced in several passages.

Page 3, Eq. (1-3) and following. The definition of the amplitude adopted in this book has not survived. A definition with a factor 1/2 inserted on the right hand side of Eq. (1-3) is now in common usage. A change in definition necessitates, of course, corrections by one or more factors of two in many locations before the results here can be compared with those in other publications.

Page 5, Eq. (1-8). The complex conjugate expression should be added to the right hand side of Eq. (1-8), which should read:

$$\frac{2N_0e^3\,|E|^2}{m^2c(\omega^2+\tau^{-2})}$$

The two sentences following this equation should be deleted.

Page 6. A factor 2 should be added to the right hand side of Eq. (1-2). This factor is due to the two permutations of the amplitudes E_1 and E_2^* in the double product of $(E_1 + E_2^*)^2$. These and other permutation degeneracy factors are carefully discussed, for example, by S.K. Kurtz in *Quantum Electronics*, ed. H. Rabin and C.L. Tang, Vol. 1A, p. 209, Academic Press, New York, 1975.

Page 7. Eqs. (1-15) and (1-18) need an additional degeneracy factor of 6 on the right hand side.

Page 28. The paragraph of lines 4-20 from the top should be replaced by the following:

> "The correct limiting behavior for the case that either the electromagnetic frequency or the material resonant frequency becomes very small, $\omega \to 0$ or $\omega_{ng} \to 0$, respectively, requires a more careful treatment of the damping terms, as has been discussed in detail by Van Vleck and Weisskopf.[3]"

Page 30. A minus sign should be added to the right hand side of Eq. (2-34).

Page 57. "ω_{ab}" should be replaced by "ω_{ba}" on the right side of Eq. (2-98).

Page 72. In Eq. (3-26) "dE" should be replaced by "–dE".

Page 73. In Eq. (3-30) "f_r" should be replaced by "f_R".

Pages 121-165. The experimental results and numerical data quoted in Chapter 5 are in need of considerable revision and updating. Since these notes were written in 1964, numerous new developments in nonlinear optics have taken place. Some of these were anticipated in the present volume, some were not.

Among the latter, we mention the phenomena of self-focusing, self-induced transparency, spontaneous parametric scattering, squeezed states, numerous forms of laser-plasma interactions including laser-induced avalanche breakdown and dissociation and ionization, above threshold ionization and other multiphoton processes.

The basic concept of the nonlinear susceptibility and the associated coupling between waves in material media has resulted in many applications and has been extended to include the interaction of light waves with polaritons, spin waves, plasmons, surface plasmons, temperature and concentration fluctuations, excitons and other material excitations. These interactions have now been studied in many different materials, including atomic and molecular vapors, liquid crystals, plasmas, large organic molecules, and new crystalline materials, and in many different geometries, including thin films, optical fibers, and quantun well heterostructures. The parametric studies of harmonic generation, sum and difference frequency generation, and three- and four-wave light mixing have been extended to the vacuum ultravioloet and to the far infrared regions of the electromagnetic spectrum.

Fortunately, several monographs on nonlinear optics are now available for the reader who wishes more up-to-date and detailed information about these developments. A limited selection for further reading follows.

1. Y.R. Shen, *The Principles of Nonlinear Optics,* Wiley, New York, 1984.

2. R.W. Boyd, *Nonlinear Optics,* Academic Press, Inc., San Diego, 1992.

3. A. Yariv, *Quantum Electronics*, Third Edition, Wiley, New York, 1989.

4. P.N. Butcher and D. Cotter, *The Elements of Nonlinear Optics*, Cambridge University Press, 1991.

5. M.D. Levenson and S.S. Kano, *Introduction to Nonlinear Laser Spectroscopy*, Academic Press, Inc., San Diego, 1988.

6. J.F. Reintjes, *Optical Parametric Processes in Liquids and Gases*, Academic Press, Inc., San Diego, 1984.

7. G.P. Agrawal, *Nonlinear Fiber Optics*, Academic Press, Inc., San Diego, 1989.